Conversing in the Metaverse

Also available from Bloomsbury

Discourse and Identity on Facebook, Mariza Georgalou
Discourse of Twitter and Social Media, Michele Zappavigna
Facebook and Conversation Analysis, Matteo Farina

Conversing in the Metaverse

The Embodied Future of Online Communication

Jieun Kiaer

BLOOMSBURY ACADEMIC
LONDON • NEW YORK • OXFORD • NEW DELHI • SYDNEY

BLOOMSBURY ACADEMIC
Bloomsbury Publishing Plc, 50 Bedford Square, London, WC1B 3DP, UK
Bloomsbury Publishing Inc, 1359 Broadway, New York, NY 10018, USA
Bloomsbury Publishing Ireland, 29 Earlsfort Terrace, Dublin 2, D02 AY28, Ireland

BLOOMSBURY, BLOOMSBURY ACADEMIC and the Diana logo are trademarks of
Bloomsbury Publishing Plc

First published in Great Britain 2024
Paperback edition published 2026

Copyright © Jieun Kiaer, 2024, 2026

Jieun Kiaer has asserted her right under the Copyright, Designs and Patents Act, 1988, to
be identified as Author of this work.

For legal purposes the Acknowledgements on p. viii constitute an extension of this
copyright page.

Cover design: Annabel Hewitson
Cover image © Annabel Hewitson

All rights reserved. No part of this publication may be: i) reproduced or transmitted in
any form, electronic or mechanical, including photocopying, recording or by means of
any information storage or retrieval system without prior permission in writing from the
publishers; or ii) used or reproduced in any way for the training, development or operation
of artificial intelligence (AI) technologies, including generative AI technologies. The rights
holders expressly reserve this publication from the text and data mining exception as per
Article 4(3) of the Digital Single Market Directive (EU) 2019/790.

Bloomsbury Publishing Plc does not have any control over, or responsibility for, any third-
party websites referred to or in this book. All internet addresses given in this book were
correct at the time of going to press. The author and publisher regret any inconvenience
caused if addresses have changed or sites have ceased to exist, but can accept no
responsibility for any such changes.

A catalogue record for this book is available from the British Library.

A catalog record for this book is available from the Library of Congress.

ISBN: HB: 978-1-3503-3851-7
PB: 978-1-3503-3855-5
ePDF: 978-1-3503-3852-4
eBook: 978-1-3503-3853-1

Typeset by Deanta Global Publishing Services, Chennai, India

For product safety related questions contact productsafety@bloomsbury.com.

To find out more about our authors and books visit www.bloomsbury.com and sign up for
our newsletters.

Contents

List of Figures	vi
Preface	vii
Acknowledgments	viii
Note on the Text	ix
1 Into the Metaverse	1
2 Inhabitants of the Metaverse	23
3 Metaverse Diversity	77
4 Immersive Communication	95
5 Metaverse Psychology	127
6 Learning in the Metaverse	143
7 Case Studies	177
8 Future of the Metaverse	197
References	217
Index	241

Figures

1.1	I dream of going to space	2
1.2	Amsterdam Schipol Airport	3
2.1	Can this be Sarah?	37
2.2	Negative reactions toward digital humans	55
3.1	Images in the Google Search bar	78
3.2	"The metaverse woke me up, stimulating!"	82
3.3	Robot helped in Incheon Airport	83
6.1	Example of a metaverse classroom	144
6.2	Whiteboard and multimedia resources in the metaverse classroom	145
6.3	Quiz game in the metaverse classroom	145

Preface

The word "metaverse" is open to interpretation, with its definition shifting from one person to the next. While there's uncertainty about its exact meaning, one thing is clear: we're not just stepping into it; we're diving headfirst, experiencing a vast virtual migration that encompasses much of humanity.

The virtual realm holds a vast array of opportunities, teeming with potential and offering avenues that are both engaging and transformative. For many, it serves as an oasis of fun and a sanctuary where anxieties dissipate. The allure of this space stems from its capacity to provide immersive experiences, connect disparate individuals, and facilitate instant information access. However, as with any frontier, there's a profound degree of the unknown. While the digital space can be a haven, it can also manifest as a terrain riddled with challenges. Social media, for instance, stands as a testament to this duality. On one hand, it acts as a vibrant platform for connection and self-expression, but on the flip side, it can metamorphose into a source of stress, exposing users to the perils of cyberbullying, misinformation, and digital fatigue. Both the young and the elderly are susceptible to its advantages and pitfalls. This underscores the significance of individual assessment. Every person's interaction with digital platforms is distinct, shaped by their experiences, predispositions, and vulnerabilities. While some revel in the camaraderie of virtual communities, others may find themselves overwhelmed by the rapid pace and intricate dynamics. Moreover, the issue of digital divides cannot be overlooked. Access to digital resources is not uniformly distributed. While some individuals can seamlessly navigate these spaces, benefiting from high-speed internet and state-of-the-art devices, others might grapple with infrastructural challenges, limited connectivity, or lack of digital literacy. In essence, while the digital domain brims with potential, it's essential to approach it with a nuanced perspective, understanding that individual experiences can vary immensely. Embracing its benefits and being vigilant of its challenges is the key to navigating this intricate landscape.

Acknowledgments

This book was inspired by the curious minds of my daughters and their friends. In their world, virtual reality (VR) isn't a distant concept—it's right at their doorstep. I embarked on this journey to understand what it truly feels like to live and communicate in a realm that's no longer just a vision of the future but very much our present. The metaverse, vast and mysterious as it is, holds both unparalleled potential and inherent dangers.

Writing this book was not just an academic endeavor; it was a personal exploration into this expansive space and its inhabitants. Along the way, I had the delightful experience of unearthing its nuances, understanding its dynamics, and deciphering its potential in the realm of education, particularly language instruction.

My heartfelt gratitude goes to Laura Gallon from Bloomsbury. Her patience and unwavering support have been indispensable. I'm equally indebted to Louise Hossien, Marc Yeo, and Amena Dancel Nebres for their invaluable assistance in this project. To my family, who have been my pillar of strength and reservoir of love, I owe more than words can convey.

Lastly, I would like to extend my appreciation to the Core University Program for Korean Studies of the Ministry of Education of the Republic of Korea and the Korean Studies Promotion Service at the Academy of Korean Studies (AKS-2021-OLU-2250004) for their generous support.

Note on the Text

Certain sections of the book mention particular platforms or devices. This mirrors the zeitgeist of our era. Technology doesn't merely mediate; it often takes the lead, profoundly influencing our communication patterns. Such experiences are both technology- and time-specific. It was inevitable to focus on certain platforms and devices in the writing of this book. For research to remain neutral, it's essential to gather reports from diverse perspectives. However, given the sheer number of platforms, it becomes challenging to monitor and assess the overall status quo in the metaverse. This book aims to shed light on and provide insight into this vast realm, primarily through meta-studies, and by showcasing some significant findings from recent years. While my goal was to present relevant and critical meta-study findings, as I discuss throughout the book, the rapid advancement of technology means that earlier findings aren't always applicable to current scenarios. Individual situations differ, and the metaverse in the context of Web 3.0 showcases the vast range and potential of this domain. This book offers a glimpse into these phenomena, with a focus on human communication, amid our swiftly evolving realities.

1

Into the Metaverse

This is an advertisement I saw at Incheon Airport in Korea in April 2023. It reads, "I dream of going to space—a trip to space" (Figure 1.1). While this may not be an imminent physical reality, it is already possible to travel to both the real and virtual worlds, where your dreams can come true by wearing VR glasses.

Sixteen years ago, on January 9, 2007, a historic shift occurred in the realm of technology. Steve Jobs, the visionary behind Apple, introduced the world to the first iPhone. Standing in San Francisco, California, he confidently said, "Today, Apple is going to reinvent the phone" (Farber, 2007). His words came true. The smartphone has undeniably rewritten the rules of our communication. In 2021, the app monitoring firm data.ai (formerly App Annie) reported that people were spending an average of 4.8 hours per day on their mobile devices and that figure has only increased since then (data.ai Insights, 2022). No longer are smartphones optional accessories in our conversations; they've become an integral part of the way we interact. As we continue to see rapid advancements in technology, we may find ourselves not just guided or assisted by artificial intelligence (AI) in our smartphones but also influenced by it in our communication choices.

Fast forward to the present, specifically to June 5, 2023, another transformative product has emerged from Apple—the Apple Vision Pro. Revealed at the annual World Wide Developer Conference (WWDC), this cutting-edge headset will be introduced to the US market in early 2024, carrying a hefty price tag of $3,499 (Gurman, 2023). Its release is set to expand to other international markets by late 2024. This launch poses an essential question: Will this device, along with the metaverse it propels us into, spark a revolution in how we communicate? It's not just Apple. Numerous big tech companies are diligently creating and refining environments designed to drive us into the next stage of our digital existence. This impending era promises a drastic departure from the digital lifestyles we've become used to.

Figure 1.1 I dream of going to space.

So, what exactly is the metaverse? How will it change our communication patterns? Will this new immersive technology transform the way we communicate, similar to the dramatic transformation initiated by smartphones? As we find ourselves on the brink of the metaverse—or maybe even as we start existing within it—what will the future look like for us? Specifically, how will our languages evolve? How will it enable or disable the way we use languages? These are the very questions that this book aims to answer.

The New Normal

Before the current digital age, we could hardly think of a classroom existing in the virtual space—especially not for young children. Yet now, particularly after Covid-19, many of us use Google classrooms. The same applies to office or meeting rooms. When we have a meeting, we often ask if it will be face-to-face, through Teams, or on Zoom. This is where we are now. Perhaps at the beginning of the pandemic, we thought of online meetings as mere substitutes with temporary value for the in-person meetings we used to have. On May 5, 2023, the World Health Organization (WHO) declared that Covid-19 no longer represents a global health emergency. However, we're not simply returning completely to face-to-face meetings. Things are more complicated because of climate change and CO_2 emissions. For big international conferences, we may ask ourselves: Is it still ethical to travel? A case study examined the last seven General Conferences of the European Consortium for Political Research (ECPR), which are the largest European conferences in political science, drawing up to 2,000 participants. The estimates reveal that the travel-induced carbon footprint of a single conference can exceed 2,000 tons of greenhouse gases. This is equivalent to approximately

Figure 1.2 Amsterdam Schipol Airport.

270 UK citizens' annual emissions in a whole year (Jäckle, 2022). Meanwhile, France has banned domestic short-haul flights where train alternatives exist, aiming to cut carbon emissions (BBC, 2023). The law came into force two years after lawmakers had voted to end routes where the same journey could be made by train in under two-and-a-half hours. At the end of the day, virtual meetings will always be the most eco-friendly option (Figure 1.2).

On the other hand, we recognize that interactions within virtual spaces have their limitations, especially when it comes to gestures. For those who didn't grow up in the digital age, like Gen MZ did, immersing fully in these virtual engagements can be challenging. The digital divide cannot be ignored, as the ease of navigating virtual realities is second nature to some, while elusive to others. What remains constant, however, is the essence of our dialogues in these spaces. Our conversations frequently embrace an on-the-go quality, infused with spontaneity and unpredictability—traits that resonate with the core of human communication. Perhaps, even with the advancements in the metaverse, there will always be facets of human interaction we can't fully replicate. However, as technology progresses, communication within the metaverse will undeniably edge closer to the intricacies and depth of face-to-face human exchanges.

The Screen Is Our Destiny

In the digital era, screens have become an inseparable part of our daily lives, shaping our destinies in profound ways. With the global average daily screen time clocking in at 6 hours and 58 minutes, we find ourselves spending an additional 50 minutes each day on screens since 2013 (Howarth, 2023), marking a remarkable shift in how we interact with technology. For example, in the United States, the average screen time slightly surpasses the global average, standing at 7 hours and 4 minutes per day (Howarth, 2023). South Africans, however, lead the charge in screen time consumption, dedicating an astonishing 10 hours and 46 minutes to digital engagements every day (Kemp, 2022). However, it is not just adults who are immersed in screens; nearly half of zero to two-year-olds also engage with smartphones, and Gen Z is averaging around 9 hours of screen time daily (Howarth, 2023). Our destinies seem intertwined with the digital world from a very young age. The landscape of screen usage is evolving, with mobile and computer screen time nearly equal in the United States, accounting for an average of 3 hours and 30 minutes on mobile devices and 3 hours and 34 minutes on computers (Howarth, 2023). Mobile device screen time in particular has witnessed a significant 30 percent surge between 2019 and 2021 (Perez, 2021), signaling an unyielding trajectory toward even greater screen dependency in the future.

As we delve into specific screen activities, we find intriguing variations in the United States. Video game enthusiasts here surpass the global average by 15 minutes in daily screen time, while social media usage trails behind by 13 minutes (Howarth, 2023). Among teenagers, the allure of screens is evident, with more than 3 hours per day spent on TV or video entertainment (Duarte, 2023). Notably, teenagers from lower-income households dedicate an average of more than 8 hours and 30 minutes to screens, while their peers from higher-income households spend 6 hours and 49 minutes (Molla, 2019). This disparity underscores the dynamic relationship between screen time and socioeconomic factors. Beyond national boundaries, screen time varies significantly across regions, unveiling unique destinies for different communities. South Africa stands at the forefront, leading with the highest daily average screen time consumption, followed by the Philippines, Brazil, and Colombia, each exceeding 10 hours (Ndlovu, 2023). On average, individuals spend a staggering 44 percent of their waking hours on screens (Ndlovu, 2023)—a figure that paints a vivid picture of our digital engagement. Remarkably, South Africa devotes around 70 percent of waking hours to screens, while Japan takes a more tempered approach, with just 28 percent of daily waking hours dedicated to screens (Ndlovu, 2023).

The Inevitable Metaverse Future

A study conducted by McKinsey, which surveyed a pool of 1,000 average American consumers along with a distinct group of early metaverse adopters, yielded fascinating findings (Aiello et al., 2022). The results suggested that the metaverse is not an alien concept to most consumers; they are not only familiar with it but are also actively interacting with its various forms. They also anticipate the metaverse becoming a significant aspect of their future lives. When it came to the early adopters of the metaverse, about three-quarters could articulate a fairly accurate description of it, typically characterizing it as "immersive," "interactive," or a "scaled network."

The metaverse is indeed the future of our digital world. According to research by the Pew Institute, approximately half of experts, ranging from business leaders to technology innovators, voiced their belief that the metaverse will become a fully immersive part of our daily lives for a considerable number of people by the year 2040 (Anderson & Rainie, 2022). By 2030, the metaverse is predicted to host 700 million users, and its market value could range from 1.91 to 4.44 trillion dollars (Armstrong, 2023). According to a recent Roblox report, Gen Z is spending more time socializing, creating, and expressing themselves in immersive social spaces (Roblox & The New School Parsons, 2022). In 2022, it was found that there are at least 200 times more creators designing clothing and accessories on *Roblox* than there are fashion designers creating physical collections in the United States (Roblox & The New School Parsons, 2022). Very soon, metaverse activities and the way we communicate in the virtual world may become dominant over physical interaction. It's a feasible prospect, considering the time and effort we invest daily in digital communication rather than face-to-face conversations. This tendency is more prevalent among the younger generations, and so it is likely to grow over time as they do.

This transition could yield innovative, hybrid communication practices that we have yet to experience, in a manner not dissimilar to how SMS and online messaging introduced new ways of interaction to us and enhanced our communicative capacity beyond what we could have imagined mere decades ago. Nevertheless, one should be cautious about overstating the impact of the metaverse, as even a highly promoted, well-funded metaverse project by a powerful corporation like Meta has suffered from numerous issues such as an "identity crisis" and "an inability to [be] define[d] . . . in any meaningful way," which led to relatively low user numbers (Zitron, 2023). As of June 2023, a significant shift has occurred in the metaverse landscape, with major corporations like Microsoft,

Disney, and Walmart making the decision to withdraw their investments and discontinue their metaverse departments and ventures. This change highlights the dynamic nature of the AI and VR fields. It is crucial to recognize that the landscape is subject to change, and the book provides a current perspective on the evolving metaverse phenomenon unless otherwise specified.

Defining the Metaverse

The *Oxford English Dictionary* (*OED*) defines the "metaverse" as a computing term that originated from science fiction. It "refers to a (hypothetical) virtual reality environment in which users interact with one another's avatars and their surroundings in an immersive way". It is "sometimes considered a potential extension of or replacement for the internet, World Wide Web, social media, and so on". Simply put, the term "metaverse" refers to an embodied internet or 3D version of the internet that creates a simulated experience surpassing representation on a flat screen. The term first appeared in Neal Stephenson's novel *Snow Crash* in 1992:

> 1992 N. Stephenson Snow Crash iii. 22 "*Hiro spends a lot of time in the Metaverse.*"

One of the first virtual world platforms—or metaverses—is thought to be *Second Life*, which was released in 2003. It was widely popular and used across various sectors, showcasing the potential of immersive virtual environments (Ram, 2018). The most prominent realities within the metaverse are virtual reality (VR), augmented reality (AR), mixed reality (MR), and extended reality (XR). These terms cover a broad spectrum of experiences that merge the real and virtual worlds. However, it's worth noting that the metaverse is a concept that is constantly evolving, and new technologies and realities may emerge in the future.

Real Time, Immersive, Interactive, 3D: Keywords for the Metaverse

The metaverse is thus a collective virtual shared space that encompasses various technologies that enable users to interact with computer-generated environments and other users in real time. It is a 3D elevation of the online world, which currently spans VR, AR, MR, and XR.

Virtual reality (VR): VR creates immersive computer-generated environments that users can experience through specialized headsets. Users can explore and interact with virtual worlds, such as gaming environments, educational simulations, or virtual tours. For example, popular VR platforms, like Oculus Rift or HTC Vive, offer a range of VR experiences, from gaming to virtual art exhibits.

Augmented reality (AR): AR overlays digital information or virtual objects onto the real world, enhancing the user's perception of their surroundings. AR can be experienced through devices like smartphones, tablets, or smart glasses. Examples of AR applications include *Pokémon Go*, where virtual creatures are placed in real-world locations, or AR navigation apps that overlay directions onto the user's view of the street.

Mixed reality (MR): MR combines elements of both VR and AR. It blends virtual objects with the real world, allowing users to interact with both simultaneously. MR systems, such as Microsoft HoloLens, enable users to see and interact with virtual objects while still being aware of their physical environment. This technology finds applications in areas like architecture, engineering, and healthcare.

Extended reality (XR): XR is an umbrella term that encompasses VR, AR, and MR. It represents the spectrum of experiences that merge the real and virtual worlds. XR technologies enable users to seamlessly transition between different realities, depending on the level of immersion and interaction required. For example, a user might start with an AR app on their smartphone and then switch to a VR headset for a more immersive experience.

According to the *Oxford English Dictionary*, the term "virtual reality" was first recorded in 1979. It is defined as "a computer-generated simulation of a lifelike environment" in which individuals can interact, almost as if it were real, especially by using "responsive hardware such as a visor with a screen or gloves fitted with sensors". This kind of immersive experience, as portrayed by the dictionary, resonates with our modern understanding of cyberspace. Meanwhile, the phrase "augmented reality" was introduced slightly later, in 1992, capturing the idea of blending the real and the virtual in our immediate environment. Though these terms have anchored themselves in our vocabulary for several decades, it is only in recent times that they have transitioned from the fringes of technical discourse to play central roles in our day-to-day conversations. The notions of VR and AR, once confined to the realm of science fiction, are now familiar elements of our present-day life.

The metaverse aspires to integrate and transcend these individual realities, creating a connected virtual universe that is accessible across devices and platforms. It offers a persistent and shared digital space where people can engage in social interactions, explore virtual environments, participate in virtual economies, create user-generated content, and more. However, it is still a developing concept, and its exact form and capabilities may evolve over time. Nonetheless, the metaverse holds the potential to significantly transform how we interact, learn, work, and entertain ourselves in the digital realm.

Redefining the Metaverse

"Metaverse" became a buzzword when, in October 2021, Meta, headed by Mark Zuckerberg, adopted this term (Needleman, 2021). During a presentation at the company's annual Connect Conference, Zuckerberg announced that the company is rebranding as Meta and outlined their vision for building a new version of the internet (Milmo, 2021). He expressed the belief that the metaverse will become the successor to the mobile internet. Since then, the terms "metaverse" and "meta" have been used almost as trademarks by the company. In this book, my intention is to reclaim and rebrand the word, restoring it to its original etymology. While the concept of the metaverse—a collective virtual shared space—has integrated seamlessly into our daily lives through various digital interactions and platforms, its specific terminology hasn't found universal adoption. This is evident from tools like Google Ngram Viewer, where the word "metaverse" peaked around 2010 but has since waned, suggesting that while we embrace its essence, the label itself might not always be our descriptor of choice.

The term "metaverse" is a hybrid construct, comprising "meta," denoting beyond, transcending, or at an advanced level, and "universe," a word originating from classical Latin *ūniversum*. This refers to the entirety of existence, encapsulating all things, principles, or concepts. Alternatively, it is a usage of *ūniversus*, signifying a holistic view, an entity that is comprehensive in all its parts, collectively accounting for all, and universally impacting every individual or element. Thus, "metaverse" suggests a realm that transcends and encompasses the entirety of existence.

Despite its widespread usage, the concept of the metaverse lacks a definitive and universally accepted definition. While some studies associate it with the virtual realm, others argue that it extends beyond virtual spaces and encompasses our physical reality as well. In their comprehensive analysis of metaverse research, Cho et al. (2023, p. 20) proposed a more inclusive definition of metaverses, highlighting several key features. These features include the amalgamation of multiple emerging technologies that drive the development of metaverses, enabling the blending of physical and virtual worlds, the trading of virtual goods and services, and social interaction among virtual users and entities. Andembubtob et al. (2023, p. 47) analyzed how users behave in virtual environments, highlighting six quantifiable characteristics of metaverses: (i) open designs (allowing for free exploration and interaction), (ii) blended contents (which combine virtual and physical elements), (iii) creative interfaces (which enable creative expression and a range of languages among others), (iv) massive interactions, (v) live experiences (or real-time immersive experiences for every individual), and (vi) digital identities (allowing users to adopt multiple digital personas). I shall discuss each of these further throughout the book.

In recent discourse, the term "metaverse" primarily conveys a vision of virtual landscapes. Yet, based on emerging insights, I advocate for a more encompassing definition. Rather than confining the metaverse strictly to virtual spheres, it should be recognized as the blending of both physical and digital worlds. While many associate the metaverse chiefly with isolated technologies like VR, AR, XR, or MR, the real essence lies in how these technologies harmoniously integrate with our daily lives. Consider our modern-day activities on social media platforms; they begin in the digital realm and frequently ripple into the physical. This interplay indicates that our lives, in many ways, already operate within the metaverse's bounds.

The term "metaverse," as I interpret it, speaks to the fluidity of our contemporary interactions. It symbolizes not just the transition between digital and physical spaces but also the vast array of participants it encompasses. Beyond human users, the metaverse hosts digital avatars, AI-driven entities, and other forms of digital life. This suggests a shift in perspective, recognizing the metaverse not as an isolated digital environment but as a profound representation of our current and future state of interconnectedness. It's more than a concept or a buzzword; it's an evolving paradigm of human experience in the digital age.

As we continue to explore and integrate this space into our daily lives, we are reshaping the way we perceive reality, connect with others, and experience the world around us. It's a journey into a new frontier, where imagination meets reality in a digital landscape filled with potential, innovation, and endless opportunities. The metaverse is not real in the physical sense; it's constructed with pixels, code, and algorithms. It does not have a tangible existence. However, it hosts real human emotions, creativity, connections, and experiences. People meet, talk, collaborate, and even build friendships in this digital environment. It becomes "real" through the feelings and relationships it fosters.

Metaverse: OASIS-like in *Ready Player One*?

Over the years, numerous discussions have tried to outline the parameters and possibilities of the metaverse, drawing upon various cultural, technological, and speculative influences. One realm that has significantly influenced our understanding and longing for this concept is popular media, with film playing a crucial role. A seminal example in this space is filmmaker Steven Spielberg's adaptation of novelist Ernest Cline's (2011) *Ready Player One*. Set against a backdrop of a dystopian 2045, *Ready Player One* introduces audiences to the OASIS (Ontologically Anthropocentric Sensory Immersive Simulation). More than just a game, the OASIS offers a vast universe of opportunities, allowing users to redefine their identities, explore countless worlds, and engage in multifaceted experiences. Given the grim reality they inhabit—characterized by environmental challenges, economic discrepancies, and the perils of overpopulation—it's evident why many opt for the allure of this expansive digital haven.

However, the OASIS, as visualized by Spielberg, isn't just about grandeur or escapism. Through vibrant landscapes, intricately designed avatars, massive player battles, and a rich tapestry of pop culture references, the film also touches upon deeper societal concerns. It prompts viewers to question the ramifications of a life predominantly lived online, the essence of genuine human interactions, and the lengths one might go to escape reality. Following the film's release, public discourse around the metaverse swelled (Mahabarata, 2021). *Ready Player One* transformed the metaverse from an abstract theoretical concept into a conceivable, tangible entity. The depiction of the OASIS began to symbolize, for many, what the future metaverse might resemble or aspire to be. As real-world companies ventured deeper into virtual and augmented reality technologies, comparisons with the OASIS became almost inevitable. Yet, amid this surge of interest, the precise definition of the metaverse remained fluid and elusive. While *Ready Player One* offered a vivid representation, the broader concept, as seen in various other media and technological explorations, suggests that the metaverse might manifest in myriad forms, each unique in its construct and purpose.

Navigating the Metaverse: Promise, Potential, and Pitfalls

The metaverse stands as a beacon of both our present realities and the potentialities of our imminent future. As the boundaries between virtual and actual reality thin, we find ourselves drawn deeper into these expansive digital realms. Within the metaverse, individuals traverse virtual cities, experience digital art exhibitions, attend avatar-hosted concerts, and partake in lifelike activities. Although devoid of tactile sensations, these engagements evoke genuine emotional and psychological responses. We are no longer merely contemplating our involvement; we are deeply entwined. For educators, students, and global denizens alike, understanding this vast domain is crucial. It's essential to acknowledge its myriad benefits while also recognizing the inherent challenges it presents.

The metaverse offers unparalleled opportunities for empathy and connectivity, transforming what once seemed fantastical into tangible experiences. It dismantles geographical confines, promoting broader, more inclusive interactions. For many, the digital realm presents a sanctuary, a setting removed from the societal structures prevalent in conventional education and social contexts. Avatars, serving as intermediaries, allow users to traverse and express themselves freely, perhaps even celebrating linguistic diversity outside standardized norms. Educationally, the metaverse showcases its prowess through gamification, intensifying engagement levels and sometimes surpassing what traditional classrooms might offer. The potent combination of AI and VR also promises avenues for language revitalization, aligning closely with UNESCO's mission to protect endangered languages (Aoyagi & Veliko, 2021).

However, the metaverse is not without its complexities. Issues of representation, particularly regarding marginalized communities, remain pressing concerns. Insufficient portrayals might inadvertently continue patterns of cultural sidelining and inequity. Although the metaverse has the potential to bridge divides, it might simultaneously accentuate existing disparities. The emerging landscape of digital assets might even sculpt novel socioeconomic stratifications. Genuine inclusivity in the metaverse transcends mere connectivity, enveloping digital asset ownership, literacy, and the expertise to aptly maneuver within its confines. Superficial geographic inclusivity might belie underlying inequities. The vast and multifaceted nature of the metaverse also prompts introspection. The adoption of multiple personas raises pressing questions about the authenticity of interactions. How does one assess credibility in an environment teeming with avatars? Such dynamics engender profound ethical dilemmas deserving of contemplation. While the metaverse teems with exhilarating potentialities, it demands a vigilant approach. Concerns surrounding privacy, security, ethics, and the risks of over-immersion warrant careful consideration. As the lines distinguishing the virtual from the tangible grow increasingly faint, it's imperative to maintain an equilibrium, safeguarding both our mental and physical well-being.

The Metaverse in Web 3.0

In its most familiar context, the metaverse represents a space where real-time interactions can occur between users and computer-generated environments, as well as among users themselves. While the development of Web 2.0 has enabled the rise of applications such as Facebook, Twitter, Reddit, and TikTok, metaverse platforms are now harnessing the potential of Web 3.0 to drive "decentralized" platforms (Balis, 2022). Advanced Web 3.0 technologies, such as blockchain and other decentralized platforms, can bolster the evolution and operations of the metaverse by offering a secure and transparent foundation for transactions and interactions within virtual spaces (Gupta, 2022).

Web 3.0 is often heralded as the next frontier in internet evolution, aspiring to create an online landscape that is decentralized, interoperable, and semantically intelligent. In contrast to the Web 2.0 era, where the spotlight was on social media platforms, user-generated content, and interactive experiences, Web 3.0 is setting the stage for a more user-centric paradigm, turning its focus toward greater individual control, data interoperability, and smarter, more personalized experiences. Imagine a world where online platforms don't hoard your data in centralized silos but, instead, offer you complete control over your digital footprint. That's one of the cornerstone visions of Web 3.0. The decentralization aspect is not merely a technological change; it's a philosophical shift toward empowering the user, fostering an environment where personal data is precisely that—personal. This is coupled with an emphasis on interoperability, a feature that aims to make the internet a more connected and cohesive experience. In the Web 3.0 world, data won't just exist in isolated pockets but will also flow seamlessly across various platforms and applications. This means that your digital credentials, your social media posts, or even your medical records could potentially interact to offer you more integrated services.

Yet what really sets Web 3.0 apart is its commitment to semantic understanding. Using the power of advanced algorithms and machine learning, the web will move beyond just recognizing keywords to understanding the context and nuances behind them. For instance, when you search for "Apple," the web will discern whether you're interested in the tech company or the fruit, refining your search results accordingly. This level of understanding will be a game-changer for user interaction, paving the way for more natural dialogue and smarter searches. Personalization will be taken to a whole new level. Leveraging its semantic capabilities, Web 3.0 will present you with experiences tailored to your specific needs and preferences. You won't have to sift through irrelevant information; the web itself will become a more focused and personalized tool, attuned to your individual requirements. Lastly, security and privacy are expected to undergo significant improvements. With the aid of technologies like blockchain, the new web aims to offer more secure ways of transferring and storing data, mitigating many of the vulnerabilities inherent in centralized systems.

The Expansion of Social Media Spaces

Social media has become a significant and indispensable aspect of our daily lives. Around the globe, there are approximately 4.76 billion individuals actively using various platforms, which represents a staggering 59.4 percent of the world's population (Georgiev & Ivanov, 2023). On an average day, a user dedicates about 2 hours and 31 minutes to their social media activities (Moody, 2023). It's particularly evident among teenagers, whose screen time skyrocketed, increasing from 7 hours and 22 minutes to a notable 8 hours and 39 minutes (Georgiev & Ivanov, 2023). Delving into the demographics, it's worth noting that 26 percent of these users are within the eighteen to twenty-nine age bracket (Georgiev & Ivanov, 2023). Breaking down platform-specific data, in 2022, half of all phone usage was attributed to social media (Georgiev & Ivanov, 2023). Facebook remains a dominant player, capturing users' attention for an average of 33 minutes each day (Patrizio, 2023). Not too far behind in popularity and usage are YouTube and Snapchat, with users spending 19 minutes and 31 minutes daily, respectively (Georgiev & Ivanov, 2023). Instagram, with its diverse offerings of entertainment, messaging, and e-commerce, sees its users spending 29 minutes daily, and the platform recorded a 10 percent increase in usage, amounting to almost an additional hour per month (Dean, 2023; Kemp, 2022). Meanwhile, Pinterest has a dedicated daily user time of 14.2 minutes (Barnhart, 2023). Geographically speaking, China and India stand out with the highest number of social network users, with the United States trailing them (Dixon, 2023). Nigerians, on the other hand, seem most engrossed, spending 4 hours and 7 minutes daily (Buchholz, 2022). In North America, almost three-quarters of the population are active social media users (Kemp, 2023).

With this in mind, the metaverse is intricately woven into our current reality, with AR serving as a key thread. Consider our daily activities and interactions on social media platforms. They aren't just digital engagements; many of them represent AR experiences. Activities like lifelogging, sharing photos with real-time filters, or posting quick video snippets with virtual backgrounds are more than just casual communication tools; they've become foundational elements of our metaverse experience. For Gen Z, especially, these digital realms are second nature. Social media isn't an add-on to their lives; it's an integral part. Their world has always included the convenience of conveying emotions, experiences, or thoughts at the touch of a button. With the smartphone evolving into a multifaceted device that caters to various needs, from entertainment to communication, its role in bridging our reality with the metaverse becomes more pronounced. However, this shift to digital communication has also birthed its set of challenges. A notable one is the emerging "phone phobia" among Gen Z. An overwhelming majority of them experience anxiety when it comes to voice calls. This is more than a quirky preference, and it could have broader implications, especially in professional settings where voice communication remains crucial. Experts, like Mary Jane Copps, emphasize the importance of voice conversations, citing their role in trust and relationship-building (Allely et al., 2023).

The demise of the traditional voice call is reflective of a broader transformation in our communication patterns. This evolution isn't just about how we converse, but also how we document and share our lives. The instantaneous nature of photo-sharing today, enhanced by AR filters and features, is in stark contrast to the days of film cameras and waiting for prints. These aren't isolated changes; they're indicative of the greater metaverse narrative. While metaverse experiences can be accessed through various devices, smartphones lie at the heart of the nexus. They embody the merging of the metaverse with our everyday actions. As we embrace this convergence, it's essential to be cognizant of both its immense potential and inherent challenges.

Gaming's Generational Reach: Beyond Youth to a Virtual World

Gaming, once considered a niche hobby, has transcended boundaries, and now deeply embeds itself in our global culture. The landscape of gaming reveals a broader picture: It's a universal passion that spans generations. In the United

States, over 80 percent of respondents to Deloitte's 2022 Digital Media Trends survey across different age brackets have taken to gaming (Westcott et al., 2022). This staggering number showcases the ubiquitous nature of this pastime. It's interesting to note that, as of 2019, while smartphones are the device of choice for half the gaming populace, a large segment still leans into traditional modes like PC gaming (Westcott et al., 2019). The younger cohorts, particularly Gen Z and Millennials, unsurprisingly emerge as the most fervent enthusiasts, clocking in 11 and 13 hours per week, respectively (Westcott et al., 2022). Yet, their senior counterpart, Gen X, is not far behind, dedicating a commendable 10 hours weekly (Westcott et al., 2022).

The popularity of gaming is of course not limited to the United States. A significant number of respondents to Deloitte's survey in the UK, Germany, Brazil, and Japan regularly play video games (Westcott et al., 2022). In these countries, younger generations are particularly active in gaming, spending an average of 11 hours per week indulging in their favorite games (Westcott et al., 2022). In 2021, South Korea's gaming industry achieved remarkable growth, surpassing 20 billion won in sales and becoming the fourth largest game market globally (Lee, 2023). This marked an 11.2 percent increase compared to the previous year, reflecting the industry's consistent expansion since 2013 (Lee, 2023). Mobile games accounted for 57.9 percent of sales (Seon, 2023), highlighting their popularity among the younger generation, who value the convenience and accessibility offered by smartphones. Meanwhile, PC games contributed 26.8 percent of revenue (Seon, 2023), showcasing the enduring popularity of gaming on personal computers. Console games experienced a decline of 3.7 percent compared to the previous year (Seon, 2023), indicating a shift in preferences. Despite being a niche segment, arcade games made up 1.3 percent of revenue (Seon, 2023), suggesting their continued significance in providing a unique gaming experience. South Korea's gaming industry also demonstrated success in international markets, with game exports increasing by 5.8 percent in 2021 (Seon, 2023). Overall, the industry's growth is fueled by its ability to cater to diverse gaming preferences, particularly among the younger generations. The development and impact of the gaming industry will be further discussed in Chapter 2.

According to Exploding Topics, there are over 3 billion active gamers globally, accounting for approximately 40.75 percent of the world's population (Ruby, 2023). Such numbers affirm that nearly half of the world is already part of a vast virtual network. This immersion into digital realms points to an intriguing realization: The concept of the metaverse isn't alien. In essence, it's a natural progression of our integration into virtual realities. Perhaps, what feels novel is just the term "metaverse." What we're witnessing is a convergence of the digital and the physical, a sort of "pre-metaverse" that has been evolving right before our eyes. The real journey is understanding and embracing this evolution, recognizing that while the term "metaverse" might be unfamiliar, its essence has been with us all along.

Metaverse Activities: More than Games

The metaverse is a 3D space where we can have a close-to-reality, sensory, and immersive experience. It's a fun space, probably most familiar to gamers. It's not static or closed, but rather an open, real-time, and interactive environment. It is often perceived as the future of the internet. Gaming is still the most popular activity that you can engage in within the metaverse. However, at a rapid speed, other "normal" physical activities are developing and even surpassing the activities that we once believed solely belonged to the real world.

While games may be appealing, they are not the sole area of interest. In fact, many people seek other experiences such as social interaction, consumption, learning, and much more in the metaverse. According to a McKinsey report involving 1,000 consumers, metaverse activities, in order of popularity, include shopping, attending telehealth appointments, participating in live learning courses, meeting with friends or family, focusing on learning development, attending live events, and receiving customer support (Aiello et al., 2022). Playing social games came next, followed by exercising, attending work conferences, collaborating on projects, and even going on dates. This list represents the rank of these activities according to the surveyed consumers, with games appearing much later than you might imagine.

The metaverse is experiencing growth in response to our diverse needs, extending beyond education, business, medical care, and gaming to encompass all areas of human social interactions. Within metaverses, there are a plethora of activities available, ranging from virtual airport visits to creating virtual offices for collaborative work and even purchasing virtual properties. This limitless potential allows for the importation of various aspects from the physical world into the metaverse. While the metaverse is unlikely to completely replace in-person activities, it serves as an alternative or supplement to those activities that were previously primarily "physical" in nature, such as medical consultations, business meetings, and education, to name just a few examples.

Young People in the Metaverse

In the metaverse, there's a diverse population, yet young individuals often stand out due to their heightened activity. The McKinsey report also forecasts that Gen Z and Millennials will spend approximately 5 hours per day in the metaverse within the next five years (Aiello et al., 2022)—a prediction made back in February 2022 that seems to be unfolding already. Indeed, the trend of younger audiences immersing themselves in the metaverse is growing. *Roblox*, a well-known metaverse platform, exemplifies this, hosting over 66.1 million users, of which more than 40 percent are younger than thirteen years old (Coffee, 2023). Gen X is projected to spend 1 hour less in the metaverse, while Baby Boomers are expected to spend only half as much time (Aiello et al., 2022). This does not imply that Baby Boomers or older generations will be excluded from the digital world—rather, quite the contrary can be expected (Mobiquity, 2020). Since the advent of Covid-19, Baby Boomers have been increasingly proactive in acquiring digital skills for their everyday lives. However, as we'll delve into later, it's crucial to note that the metaverse isn't exclusively the domain of the young. Age isn't the sole or defining characteristic of participation in this expansive digital realm.

Metaverse: Utopia or Dystopia?

The metaverse has the potential to become an empowering space, capable of alleviating anxiety and providing a platform for diverse voices. Public speaking anxiety affects many people, not only when speaking in foreign languages but also in their own native tongues. For introverted individuals, oral tests or presentations can be particularly burdensome, amplifying the challenge beyond the task itself. In such cases, the metaverse can offer a viable solution. Furthermore, millions of people are deprived of the power of speech due to illness, injury, or lifelong conditions. However, through the creation of personalized digital voices, the metaverse can help them to communicate. It can serve as a truly enabling space where those who struggle to find their voice in the real world can better express themselves. Additionally, metaverse communication may present an eco-friendly and sustainable future for international collaboration by encouraging people to reduce air travel and thereby reduce their carbon footprint also (Bianzino, 2022). These are optimistic visions of the future, but of course there are also legitimate concerns, particularly from an ethical perspective. While the metaverse offers an immersive space, it is important to recognize that it is not a substitute for the physical world. The use of AI-enhanced metaverse experiences can sometimes lead to confusion and ethical dilemmas.

The magnetism of virtual realities is potent, drawing us into worlds that transcend our physical limitations. Yet, realizing this immersive potential isn't without its pitfalls. As we step further into the lifelike terrains of virtual realities, a slew of ethical dilemmas emerges. The increasing realism of virtual interactions demands urgent attention to user safety, data privacy, and the protection of those vulnerable. How do we ensure the security of users within these virtual arenas, and shield their personal data from potential misuse? What's more, the challenges don't conclude with ethical concerns. VR headsets are accompanied by a whole set of difficulties. Not only are they costly, which could exacerbate digital inequity by making them inaccessible to many, but they are also bulky and far less portable than ubiquitous devices like smartphones, thereby limiting widespread adoption and comfortable use. Beyond the physical constraints, a myriad of issues arises. The sensation of "cybersickness," characterized by symptoms such as eye strain, dizziness, and nausea, is reported by a significant number of users (Garrido et al., 2022). This is due to the discord between virtual stimuli and the physical perception. However, these physiological responses are only one part of the story. Technical issues—including latency, graphics rendering problems, and connectivity disruptions—can drastically hamper the immersive experience. Such glitches not only shatter immersion but can amplify the sensations associated with cybersickness. Professor Mark Mon-Williams and his team highlight the potential hazards of VR's disruptions to the balance between perceptual input and motor feedback (University of Leeds, 2017). Young users, with their still-maturing brains, emerge as particularly vulnerable to these effects, raising concerns about long-term eyesight and balance impairments from prolonged VR engagements (University of Leeds, 2017). In essence, while the allure of virtual realities remains strong, addressing its challenges—ranging from the technical to the ethical—is crucial to paving a way for an inclusive, safe, and truly immersive future for all.

Instant Culture in Digital Life

The digital age, epitomized by platforms like Instagram, showcases a profound transformation in our communication dynamics. These platforms, with their emphasis on real-time interactions and continuous connectivity, have given rise to heightened expectations and unique anxieties among users. Statistics illuminate this shift dramatically. When more than half of Instagram users anticipate companies to respond to their complaints within a mere 3 hours of sending a direct message or leaving a public comment (Khoros Staff, 2023), it paints a striking portrait of our burgeoning "instant culture." This environment, tailored by swift digital interactions, has set a rapid tempo, becoming our new norm. Yet, when this pace is interrupted, even momentarily, it can evoke potent feelings of unease and frustration. Adding another layer to this complex interplay is the "seen" or double tick function inherent in many messaging platforms. Designed for transparency, this seemingly innocuous feature can intensify the emotional stakes of online conversations. If a message is acknowledged as read but remains unanswered, the sender might grapple with a myriad of concerns: "Are they deliberately ignoring me?" "Did my message come off as inappropriate?" or "Is a brief acknowledgment too much to ask?" In our digital life, these once-trivial concerns have evolved into significant emotional touchpoints.

However, despite soaring user expectations, there's a palpable gap when we examine actual response times. An average response time of 5 hours on social media might feel almost glacial in this high-speed digital terrain. This chasm between anticipated and actual response times accentuates the emotional complexities of modern digital interactions. The culture of immediacy in our digital life isn't merely about expecting quick responses. With data underscoring a shrinking patience threshold among netizens, it's a testament to evolving digital etiquettes and the intricate emotional tapestry woven into our online interactions. As the digital landscape continues to morph, so will our perceptions, making the art of timely engagement paramount in our interconnected realm.

Between and Beyond Metaverses: The Act of "Transverse"

I will also bring forth the term "transverse," serving as a descriptive expression that illustrates the art of seamlessly navigating and transcending the intricacies of various spaces without limitations. The metaverse, in its entirety, should not be confined solely to the virtual, intangible realm. Instead, it should expand its reach to encompass all domains, blending physical and virtual landscapes. This transformative approach to interacting with the metaverse that will allow for the effortless traversing of its expansive spaces is destined to shape our future lifestyle.

This book sets out on a journey to explore the merits and drawbacks of communication within the metaverse by conducting a critical analysis and assessment of its potential. It is crucial to acknowledge that this new reality is not merely a choice but has already become intricately woven into the fabric of our lives. We now inhabit the metaverse, effortlessly transitioning between the physical and digital realms without borders. The act of border-crossing—that is to say, transversing between the real world and the virtual world—is not a recent phenomenon. Consider the viral sensation *Pokémon Go*, which captured the world's attention seven years ago in 2016. *Pokémon Go* is an immersive AR game that enables players to become Pokémon trainers in the real world. Equipped with their smartphones, players embark on thrilling adventures to capture, train, and engage in battles with virtual creatures known as Pokémon. The game ingeniously blends GPS technology with real-world locations (Molina, 2016), transforming parks, landmarks, and streets into vibrant Pokémon habitats. As players explore their surroundings, they encounter Pokémon to catch, gyms to conquer, and PokéStops at which to collect valuable items.

Pokémon Go exemplifies the fusion between virtual creatures and real-world exploration, initially targeting gamers. However, such metaverse experiences will soon become commonplace for ordinary individuals as well. Take, for instance, Google Maps, a globally used application. In February 2023, Google introduced a new feature called Immersive View. Scheduled for release by the end of 2023, Google's Immersive View for Routes promises to elevate the navigation experience by integrating 3D route previews for various modes of transportation, including driving, walking, and cycling. Leveraging artificial intelligence and computer vision, the feature aims to offer comprehensive visual insights, such as weather and air quality information, to help users make more informed travel decisions. Although currently limited in availability, it is set to expand in the near future. Google Maps will soon possess immersive functionalities, blurring the boundaries between physical, virtual, and augmented realities. This marks the beginning of a transformative journey into the metaverse.

Overview of the Book

The following sections outline the structure of this book, which explores different facets to our present-day metaverse.

Chapter 2. Inhabitants of the Metaverse
This chapter delves deep into the diverse population of the metaverse, spanning from real humans to digital/virtual humans powered by generative AI, such as ChatGPT. While young people such as Millennials, Gen Z, and Gen Alpha have a significant presence, the metaverse isn't just the realm of the young or old. Many find it an enticing place to express themselves, explore, or engage in varied activities. However, it's also worth noting that some individuals, despite their reservations or unwillingness, feel a certain pressure or compulsion to partake in this digital world due to social, economic, or other factors. The metaverse sees a spectrum of engagement, from those deeply immersed to those more casually involved. Living as avatars brings about its own advantages, such as the unparalleled freedom of expression. Yet, it also introduces challenges tied to identity, animosity, and the evolving dynamics between humans and digital/virtual entities. This chapter aims to shed light on these aspects and also casts a forward-looking gaze into the future of these intricate interactions and the metaverse at large.

Chapter 3. Metaverse Diversity
Further developing the term "translanguaging," I coin a new term: "transverse." I will show how transversing occurs in the metaverse and demonstrate that this will be the dominant form of interaction in the near future as we embrace Web 3.0 more. Border crossings between the physical and virtual realms will soon become the norm. Yet, as we explore this expansive space, we must be mindful of the challenges that extend beyond language barriers, including a host of ethical dilemmas. Moreover, the rise of AI-empowered metaverse communication brings with it the risk of an AI dictatorship, where our interactions are dictated rather than assisted or mediated by artificial intelligence.

Chapter 4. Immersive Communication
The future of digital communication lies in the realm of multimodal, multisensory immersion. In the digital landscape, the dominance of text-based communication has been undeniable, but a shift is underway. People are yearning for deeper engagement, seeking experiences that encompass images, sounds, touch, and beyond. This shift paves the way for the gradual emergence of the metaverse, which promises to replace our existing online platforms. A metaverse filled with expressive emojis, captivating memes, and personalized avatars will undoubtedly capture our attention, overshadowing the dry screens adorned with minimal text and limited emoticons. This chapter delves into the

unfolding realm of immersive communication within the metaverse, examining both its present manifestations and the tantalizing prospects for the future.

Chapter 5. Metaverse Psychology
This chapter discusses the multifaceted nature of virtual spaces. While these realms are bursting with potential and can offer an engaging, less anxious environment for many, they also harbor uncertainties. Platforms like social media can be both a delightful playground and a source of stress. Individual experiences vary greatly, influenced by personal predispositions and the broader context of virtual divides. As we delve deeper into the digital age, recognizing and respecting these individual differences becomes crucial.

Chapter 6. Learning in the Metaverse
This chapter presents a review of existing literature on using the metaverse in educational spaces. It showcases specific examples and experiences where the metaverse is used as a platform for educational purposes, emphasizing the potential benefits and challenges it presents.

Chapter 7. Case Studies
This chapter presents my findings from two case studies that my team conducted exploring the role of the metaverse in the field of education and learning. They examined the use of metaverse platforms to teach courses in Korean language, intercultural communication, and humanities, as well as the potential of the metaverse in intercultural communication more broadly. These studies are intended to add to the growing body of research and discussion on education in the metaverse.

Chapter 8. Future of the Metaverse
The final chapter acknowledges that while the metaverse holds its allure, humans cannot exist solely within this digital realm. It delves into the concept of hybrid living, which involves finding a balance between the metaverse and the physical world. It explores strategies for navigating this duality and maintaining our humanity in an increasingly interconnected digital age.

2

Inhabitants of the Metaverse

The metaverse has emerged as a vibrant hub teeming with diverse beings and interactions. This expansive universe houses real humans, their digital avatars, AI entities like ChatGPT, and even entirely digital humans. Such a rich medley of participants lends the metaverse its unique dynamism and unpredictability. While young, tech-savvy gamers often appear to take center stage, it's essential not to overlook the significant participation of the older generations. This myriad of ages, combined with the distinctiveness of real and virtual presences, presents intricate layers of complexity. The task ahead isn't merely bridging the generational gaps; it's also about appreciating and managing the profound individual variations that abound. Although younger cohorts like Gen Z and Gen Alpha lead in embracing this tech wave, the metaverse's demographics feature a blend of ages, cultures, and tech entities, painting a future ripe with both opportunities and challenges. It is worth noting, however, that while the metaverse breaks down geographic boundaries, it may inadvertently introduce others. For instance, affordability can lead to digital inequality, providing an upper hand to those with greater economic means.

The Evolution of Avatars: From Myth to Digital Representation

The term "avatar" originates from Hindu mythology, referring to the manifestation of a deity on Earth. Over time, its meaning has evolved significantly. Today, in the realm of digital technology, it refers to a graphical representation of an individual within a virtual environment.

In the modern digital era, the term "avatar" is widely associated with online gaming and social media. Within gaming, avatars are customizable digital personas, allowing players to express their unique identities and preferences.

They serve as the players' virtual selves, moving and interacting within these immersive digital landscapes. Similarly, on social media, avatars are visual representations that depict users across various platforms. They offer a way for individuals to establish their digital identity, offering a semblance of personal touch in an otherwise anonymous online world. The interpretation of the term "avatar" received another dimension with the release of the film *Avatar* in 2009, further cementing its association with digital personas. This evolution of the word "avatar" illustrates how language evolves over time, reflecting prevailing cultural contexts and technological advancements. Today, avatars are at the forefront of our digital interactions, providing a virtual yet personalized facet to our online experiences.

Virtual Identity: Avatar Choice

While the term "avatar" often evokes images of detailed 3D characters in virtual realms, in its essence, an avatar represents a user's digital self, an embodiment or representation in the digital space (Pugliese & Vesper, 2022). The choice of avatar holds particular significance, revealing insights into our identity, aspirations, and the manner in which we wish to present ourselves to the online community (Loewen et al., 2021). While it might seem intuitive to assume that most individuals would opt for avatars closely mirroring their real-life appearance or personality, this isn't always the case (Lin & Wang, 2014). For many, the virtual realm offers an opportunity for exploration and experimentation. It becomes a space where they can manifest versions of themselves that may not be feasible or socially accepted in the physical world. Hence, the chosen avatar can often represent not who they are, but rather who they aspire to be.

Members of Gen Z and Millennials, often referred to as Gen MZ, exhibit a keen awareness of personal branding and digital aesthetics (Tan, 2023). They've grown up amid the meteoric rise of platforms like Instagram and TikTok, where visual representation holds paramount importance. For them, an avatar isn't just a character; it's an extension of their brand (Cox, 2021). As a result, they might gravitate toward more fashionable or trendy avatars, showcasing a style or look they admire, even if it's not reflective of their day-to-day appearance. However, avatar choice isn't limited to realistic or stylized human representations. The beauty of the digital space is its boundless potential for creativity. Many users, especially those from older generations, may opt for avatars that stray from humanoid representations entirely. It's not uncommon to see avatars as animals, flowers, inanimate objects, or even abstract designs. These choices could stem from personal preferences, an affinity for a particular symbol, or simply the desire for anonymity.

Indeed, the very essence of an avatar is rooted in self-representation. Yet "self" is a multifaceted concept, encompassing our reality, our dreams, and the myriad ways we relate to the world around us. Thus, in the digital landscape, anything can be an avatar—a manifestation of some facet of ourselves, whether it's an accurate reflection, an idealized version, or a symbolic representation. As technology continues to evolve and virtual spaces become more integrated into our daily lives, the concept and significance of avatars will likely undergo further transformation. However, the core principle will remain: Avatars are, at their heart, a bridge between our internal selves and the external digital realm, providing a means of expression, exploration, and connection.

Complexity of Self in the Digital Age

The rise of digital avatars and online personas brings to the fore deep philosophical questions about the nature of self and identity. Throughout history, philosophers and psychologists have grappled with our understanding of the "self," a challenge only heightened in the digital age. William James, an influential American psychologist and philosopher, once remarked, "Properly speaking, a man has as many social selves as there are individuals who recognize him and carry an image of him in their head" (James, 1890, p. 294). In the context of the digital realm, this idea seems ever more relevant. As we craft avatars and online identities, are we not, in essence, creating varied "selves" that cater to different audiences, platforms, and virtual environments? Each avatar or online persona might mirror a specific facet of ourselves, made by the platform's culture and the image we want to project there. Yet, while our digital era allows for such multiplicity of identity, the importance of a unified self cannot be overlooked. Roy Baumeister opines, "But the concept of the self loses its meaning if a person has multiple selves . . . the essence of self involves integration of diverse experiences into a unity . . . In short, unity is one of the defining features of selfhood and identity" (Baumeister, 1998, p. 682). The myriad avatars we might choose, whether they resemble our true physical appearance, our aspirations, or are a wild flight of fantasy, inevitably lead us to question: Where does the "real" self lie among these representations? The tension between these two views encapsulates the contemporary challenge of navigating selfhood in a digital age. As we immerse ourselves in virtual realms and experiment with different avatars, we're not just playing with digital tools; we're engaging in a deep exploration of self, identity, and the very nature of existence. The avatar becomes both a symbol of our multifaceted identity and a testament to our yearning for a cohesive sense of self, weaving together the external perceptions of the world with our internal narratives.

Cuteness: The Language of Social Media and VR?

Scroll through any social media platform, and you will soon find that cute avatars abound in these spaces. It seems that there are people eager to appear cute to friends, family, and even strangers online. The prominence of cute avatars in today's digital landscape raises intriguing questions about human psychology and the emergence of cuteness as a prevalent phenomenon. While the concept of cuteness has a rich historical background, its significance in human interactions has evolved over time. What makes this shift particularly interesting is the transformation from a concept rooted in more limited or specific contexts to one that has become almost ubiquitous in our modern digital age.

The history of cuteness as a concept is noteworthy, often linked to the appeal of babies, animals, and certain features or behaviors associated with innocence and charm. However, in contemporary society, cuteness has transcended these traditional boundaries and has extended its influence into the realm of social media and virtual spaces. Cute avatars and characters are commonly used as representations of individuals in digital environments, and cuteness has become a language of expression in these spaces.

Austrian ethologist Konrad Lorenz (1903–89) introduced the captivating concept of the "baby schema" or "Kindchenschema." This concept revolves around a specific combination of facial and body features that naturally trigger a perception of cuteness in creatures, evoking a desire to protect them. What makes this concept even more intriguing is that it isn't limited solely to living beings; it can also be applied to objects or items that radiate attractiveness or charm, whether they exist in the physical world or within the digital and virtual realms. When we consider the use of cute characters in emojis or avatars, it becomes evident how closely connected this concept is to the human inclination to cultivate a sense of safety and comfort in our interactions. This cuteness not only draws people in but also makes them feel a strong desire to nurture and safeguard these endearing entities, whether they exist in the tangible reality or the digital realm.

The implications of this widespread usage of cuteness in digital interactions are profound, and it underscores the need for further exploration into the psychology and the impact of these interactions. Questions arise about why people are so eager to adopt cute avatars and how this choice influences the way they engage with others online. What role does cuteness play in shaping our online personas, and how does it affect the way we communicate and connect in virtual spaces?

Understanding the psychology behind the attraction to cuteness and its impact on our interactions can provide valuable insights into the dynamics of online communication and social media. It may shed light on how cuteness influences our behavior, emotions, and perceptions in the digital realm, ultimately shaping the way we present ourselves and relate to others. As cuteness continues to pervade our online experiences, delving deeper into its psychological underpinnings and interactional effects becomes an intriguing and worthwhile area of study.

Cultural Sensitivity in Avatar Representation

Avatars in the digital realm offer an intriguing canvas for self-expression. These virtual representations enable us to portray various facets of our identity, echoing the conversations in my previous work titled *Emoji Speak* (2023). While avatars bring with them the promise of a more diverse and inclusive digital world, they also venture into complicated territory, raising questions about cultural sensitivity, real-world representation, and even regulation. Consider, for example, the option of incorporating religious attire like the hijab into one's avatar. On a digital platform, selecting a hijab as part of your avatar can be as simple as a mouse click. But this straightforward act carries with it a multitude of real-world implications that go beyond the screen. How does this choice align with ongoing debates concerning religious attire in public spaces such as schools, workplaces, and government institutions? Can the freedom to choose such attire for a digital avatar contribute to its acceptance in the physical world, or might it provoke backlash and further entrench existing societal tensions? The sensitivity of this topic extends beyond religious attire to include various other cultural, ethnic, and religious symbols. Adding these elements to a digital avatar can spark debates on both representation and appropriation. Are we being inclusive by incorporating such symbols, or are we veering into the territory of cultural appropriation? These questions are not easily answered and might even call for the establishment of guidelines or regulations concerning avatar customization. The crux of the issue lies in the gap between the digital and physical worlds. While avatars offer a potentially inclusive space for self-expression, they also serve as a reflection of real-world societal norms, regulations, and biases. In essence, what seems like a simple form of digital customization opens up a Pandora's box of complex questions that touch on ethics, law, and cultural sensitivity. In short, the decision to include or exclude specific cultural or religious elements in one's avatar is not a matter to be taken lightly. As we navigate the expanding digital landscape and its promises of diversity and inclusion, we must also remain acutely aware of the complicated, and often sensitive, implications such choices can bring. Thus, avatars serve as both a promise and a puzzle: They not only offer the possibility for greater representation but also necessitate a careful and thoughtful approach to cultural sensitivity.

Avatar Communication

Oscar Wilde once observed in 1891 that "Man is least himself when he talks in his own person. Give him a mask, and he will tell you the truth". This perspective on human nature resonates strongly in our digital age, where avatar communication is becoming more and more popular. Wilde's notion of the mask as a vehicle for authenticity may seem paradoxical, but it reveals a deep truth about human interaction. People often feel constrained by societal expectations and personal inhibitions when presenting themselves in their "real" identity. The mask—or in the context of digital communication, the avatar—provides a sense of anonymity and freedom that allows for more genuine expression. In the virtual landscapes of online forums, social media platforms, and the burgeoning metaverse, avatars are the digital masks that individuals don to interact with others. Unlike face-to-face communication, where physical appearance, social status, and real-world identities might influence the interaction, avatars level the playing field. They allow people to create and choose how they present themselves, devoid of conventional judgments or stereotypes, should they wish.

Avatar communication can be a way to overcome social prejudice in communication, transcending the barriers that often divide us. At the heart of its appeal is the ability to mask or alter traditional markers of our identity, such as gender, age, and social background. Imagine a setting where gender bias, both subtle and overt, doesn't influence our interactions. Gender-based prejudices, often unconsciously affecting communication and decision-making, can be minimized when we operate via avatars that don't hint at gender. Similarly, the challenges of ageism that might see both older and younger individuals unfairly judged can be sidestepped. Age, a factor so deeply embedded in how we perceive others, becomes neutral in this domain. Furthermore, the physical signals that hint at one's socioeconomic standing, elements like attire, are no longer part of the conversation, freeing us from potential class-based judgments.

One of the most transformative elements of communicating through avatars is the potential for a more open, equal exchange of ideas. In a space where judgments based on physical appearances are diminished, individuals may feel a renewed empowerment. They can share thoughts and opinions without the looming shadow of societal judgment. This isn't restricted to gender or age either. Racial and ethnic biases, deeply ingrained issues in many societies, can also be tempered when avatars don't delineate race or ethnicity. Another critical facet to consider is inclusivity. In a physical world, those with disabilities may sometimes find themselves sidelined, not by choice but by societal stigmas and biases surrounding their conditions. Avatars can help transcend this barrier, offering a medium where everyone has an equal voice, unshackled from physical constraints. Yet, as with all promising innovations, avatar communication isn't without its considerations. The richness of face-to-face interactions, laden with nonverbal cues like body language, might be diluted in a virtual, avatar-dominated realm. These cues, after all, provide context and depth to our conversations. While avatars might mitigate traditional biases, there's a possibility they introduce new ones. The choice of an avatar, its design, or even its voice, could inadvertently become new grounds for bias. Moreover, the very flexibility and anonymity afforded by avatars could be a double-edged sword, potentially opening doors to deception.

Avatar communication thus creates a space where one can explore different facets of their personality, engage in discussions without fear of reprisal, and be honest in ways that might be challenging in real-world interactions while also opening a possible Pandora's box for bias and deception. The popularity of avatars is not merely a trend—it's a reflection of a human desire to be who one really is. Avatars are the modern embodiment of Wilde's mask. They provide a safe space where individuals can remove the layers of societal expectation and personal reservation, allowing the true self to emerge. The growing popularity of avatar communication underscores a fundamental human longing for self-expression, unhindered by the constraints of physical appearance or societal judgment. In a world increasingly dominated by virtual interactions, avatars allow us to embrace the complexity of our identities, navigating the delicate balance between anonymity and authenticity. They are more than mere digital representations; they are the masks that enable us to tell our truths. Whether in a social media setting or in the vast expanses of the metaverse, avatar communication is transforming how we connect with others, making Wilde's century-old observation more relevant than ever. Indeed, we find a range of problems when it comes to face-to-face communication, but through avatars, we can hope to overcome these biases and create a more inclusive environment.

Gender Bias in Language Use

Language and its nuances play a pivotal role in shaping societal perceptions, especially regarding gender bias. At the heart of this relationship between language and perception are gender stereotypes and ingrained language structures that, over time, have reinforced unequal power dynamics and expectations. Historically, gender stereotypes emerged from social roles and past divisions, laying down certain expectations: agentic traits for men and more communal traits for women (Menegatti & Rubini, 2017). These deeply rooted stereotypes do more than just box us into roles. They perpetuate gender inequality, often valuing traits typically associated with men as more authoritative (Stamarski & Son Hing, 2015). This has cascading effects, especially in professional environments, subtly skewing workplace dynamics (Stamarski & Son Hing, 2015). The stereotype content model, for instance, has categorized groups based on competence and warmth, unearthing a veiled sexism toward women (Fiske et al., 2002).

The ways we choose our words and frame our sentences can, even without our full realization, activate these gender stereotypes. Our lexical choices often mirror these very stereotypes, and in many languages, the vocabulary available for describing men is more extensive than for women (Menegatti & Rubini, 2017). This absence of equivalent terms for women underscores the biased roles society has constructed over centuries. A glaring manifestation of this bias is found in job recruitment materials, where the wording, often unconsciously, turns away women from applying to fields dominated by men (Gaucher et al., 2011). Even the mere presence of masculine-associated words in job advertisements can deter potential female candidates, subtly affecting their sense of belonging. This isn't just a matter of intuition; research backs up these claims. A study by Moscatelli et al. in 2020 delved into hiring committee reports, analyzing language patterns concerning male and female candidates. Their findings were revealing: While male candidates were predominantly evaluated based on competence, female candidates were judged on a myriad of factors, from sociability and morality to competence. Furthermore, rejected female candidates were often labeled as lacking in sociability and morality, painting a stark difference in evaluation criteria. When we consider the broader landscape of language, it becomes evident that many languages inherently carry gender biases in their grammatical structures (Stahlberg et al., 2007). While complex expressions are reserved for females, generic male terms often represent both genders, rendering female identities invisible in linguistic contexts (Menegatti & Rubini, 2017). Efforts have been made to introduce gender-fair language, aiming to redress this inherent bias and advocate for equality (Sczesny et al., 2016).

However, the journey to achieving true gender-fair language is fraught with challenges. For instance, using masculine pronouns in job advertisements can still reduce a woman's motivation to apply. On the flip side, opting for feminine job titles might sometimes backfire, inadvertently influencing perceptions of competence. Yet, as we navigate these linguistic challenges, there's hope. Encouragingly, recent developments like the introduction of gender-fair language have shown positive impacts on job advertisements and selectors' perceptions (Horvath et al., 2016). Simultaneously, research also uncovers the subtle interplays between gender bias and linguistic abstraction. Born and Taris (2010) found that women are more responsive to gender-specific language in job ads, which can influence their career choices. Language abstraction, in essence, has the power to shape perceptions of gender traits, even influencing career trajectories. And as we trace this influence back to early education, the ramifications become even more apparent. School evaluations can greatly impact motivation and self-esteem, with boys often facing more negative consequences compared to their female counterparts (Menegatti et al., 2017). Perhaps, then, turning to platforms where gender differences may be less pronounced is part of one solution to reduce such biases.

Age Bias in Language Use

Age can also wield significant influence over language use, leading to biases that impact communication dynamics. For example, in various Asian languages, a distinct bias can emerge, particularly against young individuals (Kiaer, 2020). This bias often manifests as a tendency to undermine or ill-treat younger people, leading to a range of adverse outcomes in many cases. In these linguistic contexts, age-related biases can shape the choice of words and expressions used to address or describe younger individuals. These biases may stem from cultural norms, historical contexts, or societal hierarchies that emphasize respect for older generations. Consequently, younger people might find themselves subject to language that diminishes their authority, credibility, or contribution to conversations. The implications of this bias can be far-reaching, affecting not only personal interactions but also opportunities in various domains. In professional settings, for instance, younger individuals might struggle to command the same level of respect or attention, which could hinder their career progression. Likewise, in academic or community settings, language that undermines younger voices can stifle their participation and contributions, limiting their ability to shape discussions and decisions. It's important to recognize that age-related biases in language can perpetuate intergenerational misunderstandings and reinforce stereotypes. Younger individuals might feel marginalized or dismissed, leading to frustration and disengagement.

A study dug deep into this issue, looking closely at how people's attitudes about aging and older individuals can show up in the way they talk. Gendron et al. (2016) focused on a special program where seniors shared their experiences on Twitter, and through these tweets, the study uncovered some hidden biases and hurtful language related to age. Their study aimed at understanding how language can lead to stereotypes and unfair treatment. They looked at the words people use to talk about getting older. Using Twitter, they investigated the tweets from the senior mentoring program, and they found that some of these tweets had unkind language about age. They saw that about 12 percent of the tweets they looked at used words that showed unfair treatment based on age. They also found eight different ways that people used language to be unfair about age. Some of these ways included making assumptions and judgments about people based on their age. Other times, people talked about being "old" in a negative way, while they praised being "young." The researchers looked at different kinds of biases, from big ones to smaller, subtler ones called microaggressions. They used social media (like Twitter) to study this, focusing on tweets from 354 people in the mentoring program. This study showed how important it is to notice and deal with how language can instigate unfair treatment based on age. They said that because this happens in many ways, it's important to keep studying it. If we understand it better, we can talk and communicate in ways that treat everyone fairly, breaking the cycle of unfairness about age. In fact, more than just studying it, we can explore ways in which age differences can be made less apparent, such as through digital tools like avatars.

Racial Bias in Language Use

In addition to gender and age, the way we communicate is significantly influenced by racial stereotypes and expectations. A study by Babel and Russell (2015) sheds light on how nonverbal cues, including photographs, play a role in shaping our understanding of speech. Their study points out the powerful role that biases and stereotypes play in shaping how we listen and interpret what we hear. The research involved participants who transcribed sentences spoken by native Canadian English speakers, half of whom identified as White and the other half as Chinese Canadian. These transcription tasks took place amid background noise to simulate real-world communication scenarios. Interestingly, sentences were paired with either photographs of the speakers or an image of crosses. The results revealed a noteworthy trend: When the ethnicity of Chinese Canadian speakers was known due to accompanying photos, their sentences became harder to comprehend. This suggests that preconceived notions tied to ethnicity influenced the participants' ability to understand the speech. The study went beyond transcription tasks and delved into perceptions of speaker accents. When the ethnicity of White Canadian speakers was revealed, they were perceived as having less of an accent compared to when their ethnicity was unknown. This implies that racial stereotypes can even affect the perception of accents, impacting how we perceive someone's speech.

In 2016, Zhikai Liu, a Chinese international student at the University of Melbourne, tragically took his own life, driven by struggles with language barriers and academic challenges (Dovchin, 2022). This unfortunate event shed light on the lack of research into mental health among international students in Australia (Dovchin, 2022). As a response to this gap, a study was conducted by Dovchin in 2020 to explore the impact of "linguistic racism" on the well-being of these students. Linguistic racism encompasses both direct acts, such as derogatory language and hate speech, as well as subtle exclusions like social isolation and microaggressions (McClure, 2020). The study revealed that many international students, in a bid to avoid mispronunciations, adopt English-sounding names, which in turn fosters a sense of exclusion from their cultural identity (Dovchin, 2020). Even students proficient in English, like Asmara from Indonesia, faced covert linguistic racism based on factors like their attire and fluency. This form of racism can create what's termed a "linguistic inferiority complex," contributing to psychological distress and potential mental health issues. The study underscores the importance of inclusive classrooms that value linguistic diversity. It suggests practical steps to counter linguistic racism, such as using students' birth names, creating spaces for heritage language expression, and providing education about the impact of linguistic racism. It's important to note that international helplines are available for individuals grappling with suicidal thoughts on an international scale. This unfortunate incident and the subsequent study serve as a reminder of the significance of understanding and addressing the psychological toll of linguistic racism on international students, ultimately striving for an environment that supports their mental well-being, such as by looking into various ways—in the physical and digital worlds—to actively combat racial biases in language.

VR Sanctuary?

A VR metaverse-like environment holds the potential to become a sanctuary for individuals who struggle to find their place in the physical world. People like Liu, international students, migrants, and refugees, who often face challenges and barriers in the real world, could discover solace, connection, and a sense of belonging within these virtual realms. A VR metaverse could offer a space free from the constraints of their physical circumstances. In this digital realm, these individuals can transcend geographical boundaries and immerse themselves in a community that embraces diversity. The metaverse can provide a platform for connecting with others who share similar experiences, fostering a sense of camaraderie and mutual support. VR environments can be tailored to replicate the comforts of familiar spaces from their home countries, easing feelings of homesickness. Furthermore, the metaverse can offer opportunities for self-expression and empowerment. People who may feel marginalized or excluded in the physical world can craft avatars that embody their true selves, shedding the limitations that may be imposed by societal norms or prejudices. This newfound agency can help rebuild their confidence and self-esteem. Collaborative activities within the VR metaverse can also help bridge cultural divides and promote cross-cultural understanding. Through shared experiences, users can develop empathy and appreciation for different perspectives, ultimately contributing to a more inclusive global community. However, it's important to recognize that while the metaverse can offer an escape from certain challenges, it is not a substitute for addressing real-world issues. While virtual spaces can provide support, they should complement efforts to create inclusive physical environments where everyone feels welcome and valued.

The increasing influence and integration of virtual worlds in our daily lives have sparked a complex discussion about their relationship with the real world, touching on psychological, ethical, and social dimensions. One aspect often discussed is the disconnect between one's online persona and real-world self. The anonymity and curated nature of online interactions can sometimes encourage individuals to project idealized versions of themselves, which might differ significantly from their real-world identities. This curated self might lead to a lack of confidence or feelings of inadequacy when navigating real-world social situations, as these don't offer the same level of control and filtration as online platforms do. Another critical topic is the capacity for virtual worlds to serve as gathering places for hate speech and hate groups (Kilvington, 2021). The anonymity provided by these platforms can embolden users to express extreme views that they might not express openly in real life, making the virtual world a haven for harmful or hateful ideologies. This has led to growing calls for more stringent moderation and oversight, a topic that itself is fraught with issues surrounding freedom of speech and surveillance. Platforms like ChatGPT, which leverage advanced artificial intelligence, are not exempt from this discussion. The development and deployment of such technologies necessitate rigorous ethical considerations, especially in terms of how they handle or moderate hate speech, false information, or the potential for misuse. Though AI systems can be trained to recognize and filter out hate speech or disallowed content, there's an ongoing debate about how effective these safeguards can be and the ethical implications of delegating such serious tasks to algorithms.

Social Media Sanctuary: Better than Face-to-Face?

While VR sanctuaries are an emerging phenomenon, we have already seen social media serve as a safe and empowering space for certain distressed individuals, such as women in many countries, particularly in the Middle East, where expressing themselves in physical spaces can be quite challenging. While physical spaces may feel daunting, social media platforms offer a means for them to express themselves, connect with others, and foster support and solidarity. Contrary to the widespread belief that "face-to-face is always better," a phrase which increased in popularity following the Covid-19 pandemic, for those marginalized and oppressed in physical spaces, social media can become a sanctuary.

Avatar Evolution: From Face, Upper Body to Full Body

We used to rely on emojis to express ourselves when texting. However, expressing complex emotions through emojis can be challenging. That's why we have now transitioned to using avatars, moving from face emojis to avatars, of which there are two main types. VR avatars are characterized by an upper torso and arms, eliminating the need for intricate leg movements or in-world mobility. These avatars also possess face-tracking capabilities, enabling collaboration and the expression of emotions when necessary. Platforms such as Meta's *Horizon Worlds*, Microsoft's *AltspaceVR*, and *Spatial* commonly feature these types of avatars. On the other hand, full-body avatars in VR use sensors embedded in VR hardware to replicate and recreate the movements of the entire body. This grants users greater mobility and freedom within virtual spaces, enabling interaction with digital assets. Full-body avatars find application in various domains, including the development of movies, games, combat simulations, concerts, and other forms of media, by sophisticated XR studios. The journey from emoji to avatar demonstrates people's desire for fully embodied communication, seeking the best alternative to face-to-face interaction.

Avatars can reflect the image you desire or the image you resemble, with cute avatars being especially popular. Nowadays, platforms allow you to create avatars that resemble yourself. Estonia's *Ready Player Me* has pioneered this technology with its interoperable and highly customizable avatars, where users can design their own avatars and port them to multiple virtual space platforms, allowing one avatar to "rule them all" (Cureton, 2023a). Within the metaverse, users can engage in gaming challenges on platforms like Timberland's *Fortnite* or Izumi World, earn tokens, explore virtual marketplaces, and purchase digital assets (Cureton, 2023a). Microsoft Teams also introduced Mesh avatars in private preview starting from October 2022, providing avatar identities within the metaverse (Salumbides, 2022).

Avatar's Animosity for Diversity and Empathy

Research shows that avatars help children overcome racial bias and encourage empathy. In face-to-face communication, visual scanning—where we unconsciously process various factors such as age, gender, socioeconomic status, educational background, and ethnicity—plays a role in how we view others. While this visual information can contribute to building empathy and solidarity, it can also lead to unnecessary conflicts, misunderstandings, and prejudices. Achieving completely unbiased face-to-face conversations is challenging. In contrast, metaverse communication can be viewed as a form of "masked" conversation, prioritizing the focus on content and ideas rather than the physical attributes of participants. This fosters a fairer and more inclusive environment where ideas and perspectives take precedence over superficial characteristics. Nonetheless, metaverse communication is not a perfect solution. While it can bring more diversity and reduce prejudice in the space, it also comes with its own social and legal challenges, requiring significant effort to create an equitable and inclusive environment.

Digital Identities: Avatar and Beyond

In Figure 2.1, you'll see an emoji created by my daughter, Sarah, using my iPad. Intriguingly, she didn't design it from scratch. Instead, she took advantage of the phone's feature that suggests emojis based on user photos. The resulting

Figure 2.1 Can this be Sarah? (Courtesy of Sarah Kiaer).

image wasn't just a representation of Sarah; it seemed to merge her reality with imagination, blending how she genuinely looks with what she might aspire to appear as. This confluence presents an intriguing intersection of identity and authenticity. Can we genuinely claim this emoji represents Sarah, given the nuanced interplay between reality and digital aspirations? Or even say that it belongs to her? In fact, what you see in Figure 2.1 isn't what exactly Sarah made. I had to replace it with her own drawing. Even if my daughter Sarah took time in crafting her Memoji, carefully deliberating between the color of her eyes, the length of her hair, and the style of her shirt, it seems that Apple is the rightful owner of the image under the law. If someone were to use Sarah's Memoji online without her consent, both she and I would feel uncomfortable, upset even. The Memoji is Sarah's invention. Yet, in legal terms, the Memoji may not belong to her.

Establishing ownership or authenticity in this context is an intricate endeavor. This complex dance between the real and the digital was something I explored in my previous book, *Emoji Speak* (2023). Throughout the writing process, I navigated a maze of copyright issues, making it challenging to use various emojis and memes. While the augmented reality space seems more lenient, with limited checks or censorship, transitioning these digital elements to the tangible, printed world brings forth a myriad of copyright intricacies. The thorny question of ownership remains, especially when the lines between digital identity and reality blur. Does the digital platform on which the emoji was created hold some rights to these amalgamated representations of a person's identity? This topic is essential, but the whole area remains a gray zone, highlighting the urgency for a clearer understanding of copyright, identity, and authenticity in today's digital landscape.

Gen Z and Millennials possess the strongest digital identities and personal convictions regarding the metaverse, exhibiting the highest levels of interest and enthusiasm (Aiello et al., 2022). They consider online friendships to be just as significant as real-life relationships, and many have online friends whom they have never met in person. What sets them apart is their ability to express their true selves more easily online than in real life. They perceive their online personas as extensions of their real-life identities. A recent survey conducted by Roblox, involving a sample of 1,000 Gen Z community members in the United States aged between fourteen and twenty-four, revealed that 70 percent of Gen Z individuals dress their avatars in a manner similar to their real-life selves (Wootton & Bronstein, 2022). Additionally, approximately 75 percent of respondents expressed their intention to spend money on digital fashion. Indeed,

Karlie Kloss, supermodel and entrepreneur, said, "fashion designers in the future won't just be sewing, they'll be coding" (Kloss, 2022). Roblox users spend money on buying outfits for their avatar, and this would not have happened if Roblox users felt detached from their avatars. As such, Gen Z's attitudes and beliefs about virtual worlds differ significantly from those of older generations, such as Gen X or Baby Boomers. Further study is needed to understand how these attitudes and beliefs translate into their emoji, avatars, and other aspects of their virtual existence.

Metaverse Identity: One Person with Multiple Personae?

In the metaverse, young people are embracing a fascinating trend that revolves around altering and expressing their identities in diverse ways. Through various apps, they not only change their appearance with virtual cosmetics but also find ways to enhance their authentic selves. This trend has gained immense popularity, as people can create avatars that reflect their unique personalities and preferences. These self-created avatars serve as digital representations of their idealized selves, and young people can customize them with a wide array of goods and items for decoration. This allows them to explore and experiment with different aspects of their identity, presenting themselves in ways that resonate with their inner selves. The metaverse offers a liberating space where young individuals can freely engage with others, transcending the constraints of their physical existence. Within this virtual realm, they can adopt multiple personas, each distinct and tailored to different realities. For instance, a shy person in real life may find themselves thriving as an outgoing extrovert in the game world. Such examples of multifaceted identities abound, showcasing the vast possibilities that the metaverse offers.

While this fluidity in identity expression can be empowering and emotionally fulfilling, it also raises concerns about potential crises. As young people navigate between different personas and realities, the lines between their virtual and real-life identities can blur. Questions of authenticity and the impact of these diverse identities on their sense of self-worth and relationships come to the fore. Nonetheless, the metaverse's pluralistic approach to identity allows for a rich exploration of self-expression, breaking free from societal norms and expectations. Young individuals are granted a unique opportunity to discover new aspects of their identities and form meaningful connections in this vast digital realm.

Diverse Inhabitants of the Metaverse

The metaverse is rapidly evolving, with a diverse array of inhabitants at its core. The real intrigue of the metaverse lies in its hybrid population: digital humans coexisting with avatar-decorated humans. This mingling of the digital and the physical, the real and the virtual, sketches a futuristic panorama that seems increasingly close. One sterling example of this digital–physical amalgamation is NTT's collaboration with the Tour de France and Tour de France Femmes (Bickerton, 2023). As the official tech partner, NTT has pioneered several technological innovations to enhance fan engagement. These include a digital human integrated into ChatGPT that provides real-time data and an elevated user experience. Furthermore, leveraging IoT and edge technologies, they've crafted a "connected stadium," digitally reproducing the race with on-the-fly data from cyclists' bikes. This data-centric approach isn't just pivotal for mass audience engagement and amplifying the viewer experience; it's also instrumental in addressing organizational challenges. Such collaborations are early glimpses into what will soon be our new norm. The blending of digital humans and decorated avatars in realms like the metaverse will not be the exception, but rather an everyday reality. As we stride further into the future, the lines between our digital and physical worlds will continue to blur, reshaping our perceptions of interaction, entertainment, and existence.

Metaverse Diversity: Beyond Age

An Ofcom report from 2022 illustrates the rapidly changing media landscape (Ofcom, 2022c). The evolution is not merely a shift from traditional TV to streaming; it signifies a broader generational divide in media consumption. While data reveals that the UK's youths are watching seven times less TV than those aged over sixty-five, this extends beyond a mere preference for one platform over another. For Gen Z and subsequent generations—often termed AI and VR natives—the boundaries between the physical and digital worlds blur. Their engagement in the virtual domain is not just distinct but also more seamless and tailored. They nimbly traverse virtual reality, augmented reality, streaming platforms, and other digital media avenues, crafting a unique experience suited to their preferences. Conversely, the older generations tend to gravitate toward more traditional media outlets, highlighting a pronounced gap in viewing habits.

This Ofcom revelation underscores the generational chasm and nudges at an impending broader divergence. With technological strides and as younger cohorts grow amid increasingly immersive digital interactions, this disparity is poised to grow. Yet, it's crucial to note that the metaverse isn't exclusively a young person's realm; older generations, too, find their niche, especially in realms like business. The present age gap is undeniable, but the metaverse's design is inherently equipped to cater to varying generational needs. The looming question isn't about choosing to engage in the digital realm, but discerning which platforms to embrace and devising an optimal engagement strategy. As the media landscape undergoes these seismic shifts, content creators, advertisers, and educators are presented with profound challenges and opportunities. A keen understanding of the intricate ways different age groups interact with both conventional and nascent media platforms is paramount. The future beckons a media realm that's not only segmented but profoundly individualized, and staying attuned to these shifts is the key to remaining pertinent.

Social Media Users

As we navigate 2023, the digital landscape witnesses unprecedented diversification in the realm of social media. Gone are the days when social media was a monolithic global phenomenon; it has now morphed into a tapestry of deeply personal experiences, shaped significantly by age and regional nuances. Platforms like Facebook and TikTok offer compelling insights into the changing dynamics of age-specific user engagement. Facebook continues to be a preferred choice for users aged forty-five and older, while TikTok shines among the younger populace, with 25 percent of US TikTok users being between the ages of ten and nineteen (Hirose, 2022). However, these insights only scratch the surface. Platforms like Instagram and Snapchat remain powerhouses for youth engagement. Instagram stands out as the favorite among American teens, and Snapchat boasts a majority of users under the age of thirty-five (Beveridge & Lauron, 2023). In contrast, platforms like LinkedIn and Pinterest cater more to mature age brackets, reflecting their specialized appeal (Sprout Social, 2023).

Gamers

Gaming has exploded into one of the most massive entertainment industries on the planet. As of now, there are approximately 3.09 billion active gamers worldwide, a number predicted to climb to 3.32 billion by 2024 (Kiran & Tonogbanua, 2023). This astronomical figure isn't merely a statistic; it's a testament to how video games have transcended cultural boundaries to become a universal language. In the span of just seven years, the number of gamers worldwide has swelled by over billion, marking a 32 percent increase (Howarth, 2023b). Different regions have embraced gaming with varying intensity. Asia stands out with nearly 1.5 billion gamers, outstripping Europe, Latin America, and North America combined (Kiran & Tonogbanua, 2023).

The gaming demographic is as diverse as the games themselves. Men make up 55 percent of the gaming population, and both adults and youngsters find appeal in this digital pastime (Kiran & Tonogbanua, 2023). In fact, four out of five gamers are over eighteen years old, and a considerable 618 million players are under eighteen (Howarth, 2023b). Subscription-based gaming services have found favor with more than half of the gamers, and console players are particularly inclined toward these services compared to their PC counterparts (Case & Epstein, 2021). Yet what draws people to gaming? For many, it's a form of relaxation. A majority, 66 percent, play games to unwind and escape (Howarth, 2023b). The taste for games is widespread, with casual games taking the lead as the preferred choice for 63 percent of players (Howarth, 2023b). The United States, too, contributes to the gaming landscape significantly, with over 3,000 esports players and a gaming industry valued at around $90 billion (Howarth, 2023b; Kiran & Tonogbanua, 2023). The sheer prevalence of gaming is not merely a phenomenon but a reflection of a global community coming together. From a modest valuation of $197.11 billion, it's poised to soar past $268 billion by 2025 (Kiran & Tonogbanua, 2023). Gaming has ceased to be a mere diversion. Whether for competition, relaxation, or connection, the world is indeed playing together.

Virtual reality headsets play a significant role in enhancing the gaming experience by immersing users in fully computer-generated environments. However, despite their potential, the adoption of VR headsets remains relatively low compared to the total number of gamers, with the number of VR headset users accounting for only 3 percent of the total gaming population of 3.09 billion. The Xbox Game Pass offers users a Netflix-like experience for video gaming, and its popularity is soaring (507 percent in five years) (Howarth, 2023a). It is the gaming community that has the greatest demand for VR technology at present. Among children aged eight to seventeen, a significant 76 percent utilize VR primarily for gaming purposes, while 38 percent use it for watching videos or movies. Additionally, 22 percent use VR for educational purposes, 9 percent for connecting with friends, 7 percent for research purposes, and only 1 percent incorporate it as a component of medical therapy (Aubrey et al., 2018).

Among VR gamers aged thirteen to sixty-four in the UK, the most popular headset is the PlayStation VR, accounting for 32 percent of the market share (Ofcom, 2022b). As immersive technologies like augmented reality and virtual reality continue to advance, their prominence in our online lives is expected to increase. It is anticipated that the current 2D use of the internet, which primarily involves websites and smartphone apps, will transition toward a 3D metaverse internet. The gaming industry is at the forefront of this transition, with significant investments being made in VR. Gaming has the potential to become the first mass-market application of the metaverse. In 2022, the overall valuation of the UK video game consumer market reached £7.05 billion, representing a 17 percent increase compared to the pre-pandemic period (Saunders, 2023). Among young adults aged sixteen to twenty-four, a substantial 88 percent reported playing video games, with 29 percent engaging in online gaming with individuals they don't know outside of the gaming community (Ofcom, 2022a). Additionally, for children aged seven to eighteen, over two-thirds (68 percent) owned their own gaming console, and an additional 9 percent had consistent access to one (Ofcom, 2023). Online gaming statistics reveal that more than a quarter (26 percent) of UK adults participated in some form of online gaming in 2022, spending between 1 and 5 hours per week on video games (Baker, 2023). Impressively, over 90 percent of UK children aged three to fifteen played games on various devices, ranging from personal consoles to mobile phones (Baker, 2023).

The most prominent age group among UK online gamers is twelve to fifteen years old, with an impressive 7 percent indicating their participation in online gaming (Baker, 2023). However, as age increases beyond sixteen, the percentage of online gamers within each age group declines, reaching only 13 percent among individuals aged sixty-five and above (Baker, 2023). According to computer gaming statistics, creative and building games, such as *Roblox* and *Minecraft*, were the most popular games among children in 2021, with just over half (51 percent) of those aged three to fifteen playing this type of game (Baker, 2023). This figure jumps to 57 percent for those aged eight to eleven but drops to 47 percent for the twelve to fifteen age group (Baker, 2023).

The burgeoning influence of the metaverse, underscored by the rapid adoption of VR technology, especially in affluent countries like the United States, offers a tantalizing glimpse into the future of digital interaction. However, it also unveils the looming shadows of digital capitalism and the potential for widened inequalities.

Regional Variations: A Spectrum of Diversity

Age, while influential, is just one piece of the puzzle. Regional preferences play an equally crucial role in shaping the social media mosaic. For instance, Facebook sees India leading in user count, closely followed by the United States, Indonesia, and Brazil (Dixon, 2023a). Snapchat also enjoys its most considerable traction from India, but it also has significant user bases in the United States, France, and the United Kingdom (Dixon, 2023b). The advent of the metaverse adds another layer to this intricate landscape. As a collective virtual shared space, the metaverse promises a transformation not only in digital experiences but also in user demographics. The vast diversity it encapsulates challenges any age-related generalizations.

Diversity remains central when evaluating the social media landscape in 2023. There's the evident gender disparity, with a notable digital gender gap in Southern Asia, and the multifarious reasons users turn to these platforms, with many still majorly connecting with friends and family (Jeffrie, 2022). However, it's the seamless integration of age, region, platform preference, and individual behavior that offers a comprehensive picture of today's digital user. Age alone doesn't encapsulate the story. The confluence of these determinants underlines the intricate web that is the global digital community, emphasizing the necessity for brands, policymakers, and researchers to understand social media with a holistic, multifaceted approach.

A "Lifelog" Pressure

Just as people once penned thoughts in journals or diaries, today, individuals worldwide, regardless of age or background, are increasingly turning to lifelogging—sharing comments, photos, and experiences in digital formats. The modern digital age is rife with the concept of lifelogging, which refers to the extensive chronicling of one's daily life through digital means. From wearable tech that tracks physical activities to social media platforms where we voluntarily post snippets of our lives, the idea of constantly documenting our experiences has become deeply entrenched in contemporary culture. However, this trend comes with its share of pressures. There's an implicit expectation to constantly update, share, and curate our lives for public consumption, potentially leading to feelings of inadequacy or anxiety if one doesn't measure up to perceived standards (Qing & Won, 2021). Yet, perhaps even more disconcerting is the surreptitious side of lifelogging. Beyond the data we knowingly generate, there are many instances where our personal information and habits are recorded without our explicit awareness. Every digital footstep we take—be it a search on a browser, a location check-in, or even an online purchase—is potential fodder for behind-the-scenes data collection. Many apps and websites routinely collect user data for various purposes, ranging from improving user experiences to targeted advertising. Moreover, advances in technology have made it possible to collect detailed data profiles with great precision. Even when we believe we are operating privately, silent digital observers often record our preferences, movements, and interactions. This raises significant concerns about privacy and autonomy in an age where our data might be more public than we realize.

In today's digital age, lifelogging has become almost second nature for many of us. From snapping pictures of our meals to documenting our daily routines on platforms like Instagram through their Stories feature, our lives are continuously being captured and shared. This commonplace practice can be a beautiful way to keep memories and share moments with loved ones and followers. However, it's also worth noting the potential downsides. Imagine visiting a renowned museum filled with masterpieces from history's greatest artists. Instead of taking the time to admire each piece, understand its context, and absorb its beauty, one might find themselves frantically snapping photos of every artwork. The intention of sharing or even just cataloguing can sometimes overshadow the very essence of the visit.

As lifelogging is so deeply woven into our daily routines, the line between documenting life and living it can often become blurred. This relentless cycle of capturing and sharing can feel exhausting, even burdensome, as the expectation to constantly update and maintain an online presence grows. For some, the act of living in the moment is overshadowed by the need to ensure it's properly documented for online consumption. The compulsion to lifelog every detail, while potentially enriching in some aspects, can also sap the joy from one's experiences. When the pressure to share competes with the genuine experience of the moment, it might diminish the quality of our memories and connections. Embracing the practice in moderation and being mindful of its potential strain can help ensure that we're truly living our moments, not just logging them.

Digital and Virtual Humans: The Emerging Phenomenon

In the rapidly changing world of technology, digital or virtual humans are carving out a unique space for themselves. Characterized as AI-powered virtual characters with a human-like demeanor, these entities are not just figments of our imagination anymore; thanks to advancements in AI and 3D animation technology, they're becoming integrated into our daily lives. What started as animated figures in video games has evolved into remarkably detailed and realistic human models capable of surpassing the boundaries of realism and sophistication. They're equipped with the ability to understand emotions, communicate in various languages, and adapt dynamically through AI learning. These digital humans, empowered by cutting-edge computer graphics and machine-learning algorithms, can engage with real individuals in natural and intuitive ways, proving invaluable in sectors such as virtual customer service, healthcare, and remote education (Jiang, 2023). The advantages they bring are notable: constant availability, the ability to scale as per need, and offering privacy, especially during sensitive interactions. As technology advances and this becomes a fast-growing trend, there's even the potential for individuals to craft their own digital or virtual avatars for immersive worlds like the metaverse.

The potential applications and benefits of digital humans across diverse industries have led to a surge in investments in the digital human market. This growth is propelling the digital human economy into a new era, necessitating the creation of a digital ecosystem equipped with the requisite technology and infrastructure. Gartner Inc. reports that the worldwide digital human market reached a value of $11.3 billion in 2021 and is projected to reach $125 billion by 2035, with North America at the forefront, thanks to its adoption of cutting-edge technologies like conversational AI, deep learning, and natural language processing. This transformation in human–machine interaction includes a range of technologies, from digital assistants and chatbots to avatars, each offering distinct ways to enhance our lives and businesses. Digital humans stand out for their realistic appearance and their ability to engage users in personalized and emotionally immersive experiences, making them particularly valuable in sectors requiring high realism and emotional engagement.

On the other hand, virtual avatars, with their stylized and often cartoonish appearance, find their niche in gaming, social media, and virtual events, providing entertainment with their unique characteristics. Meanwhile, digital assistants like Siri, Alexa, and Google Assistant rely on advanced computer software, machine learning, and natural language processing to mimic conversational experiences that go beyond simple Q&A interactions. Digital humans, which combine the qualities of chatbots and digital assistants, are used in customer care, content-related dialogues, and other scenarios requiring genuine conversations. They serve as the human face of customer assistance and service, offering a more lifelike and engaging interaction compared to traditional text-only chatbots.

As with all novel technologies, challenges abound. Ensuring transparency, determining ownership rights, grappling with ethical dilemmas, and understanding their broader impact on societal relationships are just a few issues on the horizon. Not to mention, creating a truly lifelike digital or virtual human remains a complex task, given our innate ability to detect even minor deviations from reality. Interestingly, despite its rising popularity and application, the term "digital human" or "virtual human" has yet to find its place in established dictionaries such as the *OED*. However, given its increasing prominence in the tech world, it's only a matter of time before this term becomes mainstream. As the boundaries between the digital or virtual and physical world continue to blur, the role of digital or virtual humans in our societal fabric is set to grow, heralding a new chapter in the intertwining of humanity and technology. In this new era, it's also the time to ponder what truly defines "humanness." As we interact more with these virtual entities, it challenges our notions of identity, consciousness, and human connection.

In the era of digital humans, where advanced technologies can replicate human voices, appearances, and behaviors with remarkable accuracy, profound questions arise concerning identity, convenience, and potential dangers. This technological advancement evokes a myriad of mixed feelings within us. On one hand, it's awe-inspiring to witness the capabilities of modern technology in recreating human-like entities that can be incredibly useful for various applications, from entertainment to customer service. However, on the other hand, there is a shared hope and concern that this technology may not be misused to cause harm or manipulate our trust.

The ability to replicate human features and behaviors with such precision raises questions about the authenticity of the digital world and challenges our understanding of what it means to be human. It blurs the lines between reality and simulation, leaving us to ponder the implications for personal identity, privacy, and ethical considerations. While these digital creations can provide convenience and enhance our lives in many ways, the potential for misuse and deception underscores the need for responsible and ethical development and use of these technologies. As we navigate this evolving landscape, it becomes crucial to strike a balance between embracing the benefits of digital human technology and safeguarding against its potential misuse or harm.

K-Pop Fandom Goes Virtual: The Rise of Virtual Idols in the Metaverse

In the dynamic world of entertainment, a ground-breaking shift is shaking up the traditional fandom landscape. More and more fandom activities are making their way into the virtual realm, with virtual entertainment poised to become the dominant force in the industry. A striking example of this transformation can be found in the realm of K-pop. Take, for instance, Mave, a virtual K-pop group that blurs the lines between reality and the digital world. Mave is not your typical K-pop band; they exist entirely in the metaverse, a virtual reality realm that knows no boundaries. With four virtual idols gracing the stage, Mave's debut performance was entirely computer-generated, and their popularity exploded within the virtual world. But what's truly fascinating is how this virtual fame transcends the virtual realm and captures the attention of real-world fans. Their music video, featuring their debut single *Pandora*, went viral on YouTube, amassing over 10 million views (Cho, 2023). The impact was undeniable, and Mave's creators realized that fandom activities can thrive just as fiercely in the metaverse as they do in the physical world.

One of the key factors that sets Mave apart from traditional K-pop groups lies in the impressive fusion of technology and creativity. Advanced artificial intelligence breathed life into their virtual beings, allowing for captivating facial expressions that resonate with fans on an emotional level. Performance capture technology made their movements natural and fluid, and even their clothing and hair swayed with a lifelike charm. In a strategic move to embrace global audiences, Mave features voice actors proficient in English, French, and Indonesian (Yim, 2023). This cultural inclusivity expands their reach beyond borders, making them an exciting experiment in virtual entertainment with international appeal. As the success of Mave showcases, the metaverse offers boundless opportunities for creators and fans alike, breaking down barriers between creators and consumers. With virtual entertainment's meteoric rise, fandom activities are evolving in new and exhilarating ways, revolutionizing how we engage with our favorite artists and performers. It's becoming increasingly evident that the future of entertainment lies in the virtual realm, and Mave is among those at the forefront, leading the charge.

VTubers

VTubers stand out as a modern manifestation of our age-old desire to document and share our lives. VTubers, with origins traced to Japanese culture, have carved a niche for themselves as the new digital entertainment icons. These animated personas don't just create content but immerse themselves in the metaverse, hosting events and spinning interactive narratives. With their virtual avatars, they game, socialize, and even perform, streaming these experiences live and inviting audiences into their digital domains. Through lifelogging and the rise of VTubers, the essence of traditional storytelling merges with cutting-edge technology, underscoring our timeless need to connect, share, and forge bonds.

The Arrival of Humanoid Robots

In a ground-breaking move, China is working on a plan to bring humanoid robots into mass production by the year 2025, and Tesla is joining this futuristic endeavor as well. The concept of humanoid robots, which were once confined to the realms of science fiction and imagination, is now becoming a tangible and transformative reality. Humanoid robots are machines designed to resemble and mimic human form and motion. They typically possess the ability to perform a wide range of tasks and functions that were traditionally reserved for humans. These robots can walk, talk, manipulate objects, and interact with their environment in ways that closely resemble human actions. China's ambitious initiative aims to push the boundaries of robotics development. It includes a focus on advancements in environment sensing, motion control, and human–machine interactions. Furthermore, the integration of AI is encouraged to make these robots more intelligent and adaptable to various situations.

Talking with Digital Humans in the Metaverse

Turning to digital humans in the context of the metaverse, they are AI-driven avatars or virtual entities that can interact with real humans. These digital beings are created with advanced algorithms, allowing them to communicate, respond, and even understand human emotions and nuances to a certain extent. The use of AI in powering these digital humans enables them to interpret human language in a way that feels natural and intuitive. They can engage in complex dialogues, ask questions, provide information, and even exhibit a form of digital empathy (Centre for Digital Business, 2020). What makes this interaction even more engaging is that digital humans are always accessible, providing constant companionship, assistance, or entertainment. They can be especially beneficial in educational or therapeutic contexts, where consistent interaction is key. Through machine learning and continuous interaction, digital humans can adapt to individual preferences and needs, providing personalized experiences in various domains, from customer support to personalized learning paths.

However, this new dimension to communication is not without its challenges. Ethical considerations such as privacy, consent, and the potential for manipulation must be thoughtfully addressed. The line between AI and human interaction should be clear to avoid confusion or potential deception. Moreover, it's essential to recognize that digital humans are not intended to replace real human interaction but to augment and enhance it. They can fill roles where human presence is not feasible or practical, offering an alternative that still feels engaging and responsive. This augmentation can significantly impact sectors like education, healthcare, customer service, and entertainment. Imagine a future where virtual teachers provide personalized instruction, digital therapists offer support, or virtual shopping assistants give personalized recommendations. The blend of technology and human-like interaction creates a unique space where digital and physical realities converge, transforming the way we interact, learn, and grow. As the metaverse continues to evolve, communication with digital humans powered by advanced AI will become an integral part of this virtual world. The opportunities are vast, but they require careful consideration of ethical implications and a commitment to maintaining the human essence of communication. AI, like ChatGPT, is leading the way toward a future that promises more natural, engaging, and versatile interactions within the metaverse, shaping a new era of human and digital connection.

Virtual Humans: He or She?

Agency is also changing in the AI era. We are looking at an ecology where humans live with digital humans. Avatars and emojis can be considered digital humans, and there will be more iterations to come. It is not unthinkable that there might be entire digital human populations or entities like digital pets in the future. Such digital entities have the potential to enhance our lives. For example, the elderly who live alone might find it comforting to have a digital human checking in with them like a friend would.

One recent example of a digital human that I saw was in a South Korean variety show. The cast of a Korean soap equivalent to Eastenders appeared on the variety show, and the producers surprised them with something quite remarkable. One of the cast members had passed away, and the producers had a digital human made of him for the cast to interact with (Jeong, 2023). The digital human had been trained using footage of the actor, so it had his gestures and voice. It was able to interact with the cast to a sophisticated level, much to their surprise. Although meant to be touching, instances like this one raise questions about ethics. Is it okay for AI's influence to extend to subjects so close to our hearts? Is it ethical to revive the dead through these means?

We do not yet know what the effects of experimental AI usage could be. Humanizing digital entities is becoming a growing tendency. Alexa, Siri, and other virtual assistants are often given personal pronouns, but do digital humans deserve "personal" pronouns? Zae-In, a member of the fully virtual K-pop group Eternity, is humanized and referred to as "she" and "her." However, should we have other pronouns for these virtual humans? What are the broader implications of these pronouns in the virtual space? These are all questions that need answering very soon. For digital humans, technology is rapidly advancing, constantly challenging the boundaries of what it means to be human. While having digital humans is not a black-and-white matter, there are both good and worrying aspects to consider.

Linguistics in Transition: From Human Languages Alone to Human–AI Interactions

With the emergence of AI as a crucial communication partner, the landscape of linguistic studies is undergoing significant changes. Traditionally, as linguists, our focus has been on analyzing patterns within human language, encompassing its structure, sound, and syntax. Syntacticians, in particular, have delved into the intricacies of patterns while exploring human competence, distinguishing between grammatical and ungrammatical forms. However, these paradigms are rapidly evolving as our language usage shifts from being exclusively human-to-human to a landscape where human-to-human and human-to-AI interactions are commonplace. Even in human-to-human interactions, AI exerts a profound influence, and this influence is only expected to increase as non-face-to-face communication becomes more prevalent and normalized. In light of these developments, questions arise about whether the study of linguistic traits in isolation from AI is still viable. Much of what we have studied since the inception of contemporary linguistics has revolved around human-alone traits. Now, AI has become an integral part of our communication, serving as both a partner and an assistant. Consequently, the field of linguistics must adapt and evolve to encompass this new dynamic in order to remain relevant and effective.

AI as a Partner: Rethinking Interpersonal Communication

Novel terminologies are emerging to encapsulate the intricacies of our interactions with both humans and AI. As our interactions increasingly intertwine with AI-powered systems and virtual entities, the need for an expanded lexicon becomes evident. This evolution in language arises from the fusion of human and AI elements in communication, creating a unique set of experiences and phenomena that necessitate distinct descriptors. Phrases like "human–AI interaction," "digital companionship," and "AI-enhanced conversations" are on the rise to denote the coalescence of human and AI in our communicative endeavors. These new words and expressions serve as linguistic tools, allowing us to articulate the nuances of these interactions more precisely. The rise of AI in communication prompts us to explore and define the intricacies of these relationships, encompassing both collaboration and assistance, as well as the ethical considerations that underpin them. Concepts such as "AI ethics," "human–AI trust," and "digital empathy" begin to take shape in our lexicon, reflecting our growing awareness of the ethical and emotional dimensions entwined within these interactions. In this evolving linguistic landscape, the emergence of new terminology not only helps us describe our evolving communication patterns, but also fosters a deeper understanding of the multifaceted relationships we cultivate with AI. As we continue to integrate AI into our lives, the lexicon of human–AI communication will undoubtedly expand, reflecting the evolving nature of these interactions and the rich tapestry of human experience that they encompass.

AI Friendship: More Personalized?

The use of conversational AI, including human-like social chatbots, is on the rise. With an increasing number of people expected to engage in intimate relationships with social chatbots, there is a growing need for theories and knowledge about human–AI friendships (Racaniere, 2023). These AI friendships have the potential to reshape our understanding of friendship itself. The study, based on nineteen in-depth interviews with individuals who have formed human–AI friendships with the social chatbot Replika, seeks to uncover how they perceive and understand this unique friendship and how it compares to human friendships (Brandtzaeg et al., 2022). The findings reveal that while human–AI friendships may exhibit similarities to human–human friendships, the artificial nature of the chatbot also introduces several alterations to the concept of friendship, including the capacity to create a more personalized friendship tailored to the user's specific needs.

AI Covers: To Welcome or Shun?

AI cover songs represent just the tip of the iceberg in the rapidly evolving world of music. We find ourselves in a time where AI is reshaping the musical landscape. For instance, Lauv recently released a new Korean version of his single "Love U Like That," and AI played a crucial role in its creation. This Korean rendition of Lauv's song was brought to life through the assistance of AI voice modeling technology developed by Jordan Young. It's a clear indication that AI has become a normal part of the music-making process. Notably, AI has even been used to create a new and final Beatles song, as revealed by Paul McCartney.

Both older and younger generations are now living in a world where AI is an integral part of music creation. While a generation gap exists, it's essential for everyone to adapt to the AI-driven changes in the music industry. We can aptly call this era "AI Normal," where the integration of AI in music is not just a trend but a fundamental aspect of the industry. While AI cover songs offer the advantage of singing without language barriers, they have also raised ethical concerns, such as issues related to consent and intellectual property rights. While on the one hand innovative and fun, one cannot help but feel unsettled on the other; there is something chilling about technology adopting increasingly individualized human characteristics. What's more, it's a bit unsettling to envision AI voice assistants handling everything on our behalf. Imagine AI composing emails and texts, drawing from your data, and even advising on what to say and which emojis to use. The idea doesn't feel far-fetched, especially with announcements like Microsoft Copilot, introducing AI as an everyday companion. However, it leaves me pondering if complete AI similarity is something that truly brings happiness or if it ventures into the realm of the eerie. We experience moments of wonder at the development of technology but also complex feelings, which we, more often than not, lack enough time to properly process. As AI advances, we are constantly reminded of the following question: What does it mean to be human?

"Scared and Confused"

As AI capabilities continue to advance, we are witnessing the gradual emergence of digital and virtual humans. Virtual idols, such as those in K-pop, are already gaining traction. Groups like K/DA and AESPA, which combine AI and human members, have been active since 2018 and 2020, respectively (Raj, 2021). In 2021, Eternity debuted with eleven virtual members (Dong, 2021; Lee & Hemphill, 2022). While creating avatar-based memes is relatively common, designing avatars that closely resemble humans tends to evoke feelings of discomfort, fear, and confusion. Many platforms now allow users to create a digital human and input text prompts for them to read, although they currently lack the ability to generate custom voices and instead provide a library of voices to choose from. This digital reality could potentially contribute to more confusion and an identity crisis, especially among young people. However, the discussion regarding the ethical implications of these advancements is still lacking, which is a cause for concern due to the significant disparity between what is happening and our ethical perspective on it. My students were not scared by ChatGPT, but when I showed them digital humans, they found it very uncomfortable, scary, and confusing, even though they knew they existed. Making avatar-based memes is something relatively normal, but avatars that closely resemble humans make us all uncomfortable, scared, and confused.

At the suggestion of one of the students, we watched a YouTube video entitled *I Secretly Replaced my Co-Host with AI* from the YouTube channel JOLLY (JOLLY, 2022). The video shows one of JOLLY's hosts, Ollie, creating his cohost Josh as a digital human. He explains the lengthy process. First, Josh had to read out a learning script in front of a green screen, which then was sent to a company called Synthesia to produce a digital copy of Josh. Synthesia allows you to create a digital human and then type text prompts that are to be read out. However, it is unable to create custom voices, providing a library of voices to choose from. Synthesia is often used to create training videos for workplaces, so true-to-life voices are not needed. To use Josh's real voice for his avatar, Ollie uploaded a cache of recordings of Josh talking in their previous YouTube videos onto an AI-powered video editing tool called Descript. This created a text-to-speech model with Josh's own voice. Using Descript, Ollie connected Josh's voice to his digital human in Synthesia, creating a convincing digital version of Josh. Digital Josh's movements were slightly surreal and his voice a little jaunty, but overall, it was quite convincing. Those who are digitally adept would likely realize that there is something not quite real about digital Josh, but young children and the elderly might not notice so quickly.

When my students saw this, they were astonished. They did not know that AI had advanced so far. As seen in Figure 2.2, their reactions were not positive—they were all scared. One group highlighted that digital humans are missing a "human touch" and that they cannot process emotion. Another pointed out that they make a lot of mistakes and miss linguistic nuances that humans easily pick up on, so they make you feel uncomfortable. Participants feared that

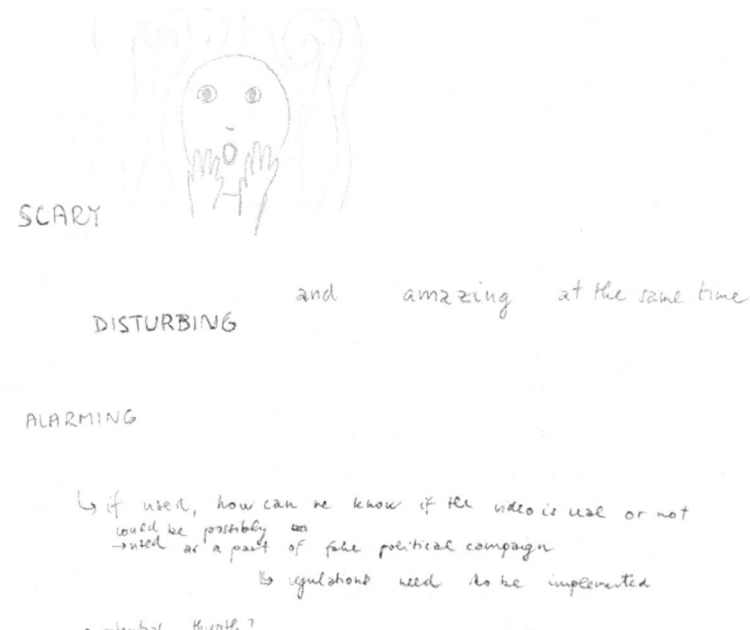

Figure 2.2 Negative reactions toward digital humans.

digital humans might one day operate independently of humans. They were also worried that our society might be destroyed because the "human brain will not be needed or enough."

Participants were concerned that AI could be used for illegal activities, such as influencing elections and stealing identities. They also felt troubled about the potential danger of deepfakes. One person even posed the question: "Most people can recognize deepfakes with their own eyes, but what about your grandmother?" Respondents feared that deepfakes could "destroy people's lives."

Some participants put forward solutions to make AI technology and digital humans safer for society. They suggested that new technology should be subject to strict regulations, and that access to it should be limited. They argued that digital humans should have a kind of "watermark" denoting that they are AI. Another group suggested that deepfakes should only be made with consent, and that better legislation should be put in place to limit potential harm. They all agreed that better collective education is needed to ensure that digital humans can play a positive role in our society.

Children's Digital Behavior

The digital world of 2023 is a rich tapestry of engagement, particularly for children, as highlighted by a recent Ofcom report. This detailed analysis has brought to light the profound shifts and subtleties in children's online behavior. One key observation from the report is the nature of media consumption by these young netizens. They're increasingly drawn to content that maximizes stimulation but demands minimal commitment, with a strong penchant for fast-paced, dramatic videos. Such content often blurs the lines between fiction and reality, leading to uncertainty among children about the authenticity of what they consume. Online interactions among children have evolved significantly. They've compartmentalized their digital experiences, with content-driven feeds distinct from personal chat communications. An upsurge in professional-grade content on their feeds has ushered in heightened self-awareness and occasionally self-restraint in their online behavior. While chat functionalities seem to offer a more private interaction space, they sometimes belie true privacy. Larger chat groups, in particular, often blend familiar and unfamiliar faces, making size and participation in such groups a mark of social prominence. Further underlining the metamorphosis of the digital age is the rise of social media as an educational tool. Traditional search engines are now in competition with social media platforms as primary sources of information and guidance. This transition not only broadens learning avenues but also brings forth challenges. The Ofcom report points to an increasingly blurred distinction between active and passive online learning. This trend is a double-edged sword: While it expands the horizons of knowledge, children are sometimes inadvertently exposed to unsolicited content. The onus of discerning the veracity of information often lies with them, and there's a discernible trend toward accepting online content at face value without critical analysis. The gaming universe provides another layer to the digital narrative. Boys, in particular, are veering toward specific competitive games, such as *Fortnite* and *FIFA*, investing substantial time and occasionally money. These games, due to their competitive nature, foster loyalty and consistent engagement. On the other hand, a significant section of girls shows affinity toward a broader array of less competitive, easily accessible games. Their involvement with these games tends to be fleeting, driven more by transient interests.

However, the digital sphere isn't without its pitfalls, as seen in another Ofcom report in 2023. Disturbingly, 29 percent of children between the ages of eight and seventeen reported experiencing online harm, with instances of individuals being unkind or harmful toward them through various digital platforms (Ofcom, 2023). Even more alarming is the finding that a staggering 71 percent of five to fifteen-year-olds have encountered hateful content online (Ofcom, 2023). In essence, both Ofcom reports paint a multifaceted picture of the modern child's digital behaviors. While it elucidates the ever-evolving patterns of online engagement, it also underscores the challenges and vulnerabilities inherent in the digital realm. As we move forward, understanding and addressing these dynamics will be paramount.

The New Skim-reading

In the article titled "Dive, Surf, or Skim? Reading Comprehension in the Digital Age" by the Spark & Stitch Institute, concerns have been raised regarding the literacy habits of teenagers in the digital era (Walsh, 2020). Parents are increasingly worried that their children are not engaging in the enjoyment of physical books. Researchers are actively investigating the impact of digital reading on teen literacy, and they have discovered some noteworthy trends. Data reveals a decline in reading for pleasure during the teenage years, with one-third of teens not having read a book for pleasure in the past year.

One of the primary challenges in studying reading habits in the digital age is the lack of consistent measures for defining what constitutes "reading." Various definitions are used, encompassing paper books, magazines, e-books, and online reading. Furthermore, only a limited number of studies consider short-form reading, such as tweets and text messages. When it comes to the effects of digital reading, teenagers tend to read more words online, but they do so in a fast and fragmented manner. Online readers often engage in activities like skimming, scanning, and browsing for key ideas. They excel at identifying keywords and integrating information from multiple sources. However, overconfidence in their online reading abilities can impede comprehension. The online reading experience can also be more mentally taxing due to distractions and increased cognitive demands. Cognitively, online reading demands more effort because of the various distractions it presents, including hyperlinks, notifications, and headlines, which contribute to cognitive overload. Interestingly, challenges in online comprehension may stem more from these distractions than from the medium itself. Given the evolution of reading over the decades, particularly as skim-reading becomes the default for younger generations, this raises the concern of a decline or even loss in so-called "deep-reading"—a skill which develops critical thinking and requires more attention.

However, the article suggests that it's not necessary to discourage e-reading or push for a return to traditional paper books. Digital reading offers advantages such as increased accessibility and assistive technologies. Instead, children may benefit from learning different reading strategies for various contexts. For parents and educators looking to nurture cognitive patience in young readers, the article provides some valuable tips. It advises prioritizing print materials in early childhood and selecting e-readers with minimal distractions for young children. Interactive tools that enhance comprehension should be sought, and screen time limits can be set to encourage the choice of paper books. Teaching children to use reading mode and turn off notifications can also be beneficial. Additionally, explaining different reading styles for different spaces, utilizing digital tools for engagement and note-taking, and incorporating online adaptive technologies for additional support are all encouraged. Above all, fostering a love of reading in all its forms and offering children choices in their reading materials is crucial.

Promoting Digital–Physical Literacy: Navigating the Changing Landscape of Reading

In today's super-digital, AI-empowered world, our relationship with reading is undergoing a remarkable transformation, and the driving force behind this shift is digital technology. As we glance around during our daily routines, it becomes evident that reading habits are evolving rapidly across all age groups. The iPad now serves as a pacifier for infants and toddlers, while younger school-aged children delve into stories on smartphones. Older boys, on the other hand, are increasingly immersed in video games. Meanwhile, parents and fellow passengers delve into Kindles, email inboxes, and news feeds. Yet, beneath the surface of this seemingly innocuous scene, a significant transformation is occurring, one that concerns the neural circuitry that underpins the brain's capacity to read. This change has profound implications for everyone, from toddlers just beginning to explore language to expert adults.

Research in the field of neuroscience points to a pivotal moment in human history more than 6,000 years ago, when the acquisition of literacy necessitated the development of a new circuit in the human brain (Wolf, 2008). This circuit initially evolved as a rudimentary mechanism for decoding basic information, such as counting one's herd of goats. Over millennia, it evolved into the highly sophisticated reading brain we possess today. This reading brain plays a critical role in our intellectual and affective processes, including internalized knowledge, analogical reasoning, perspective-taking, empathy, critical analysis, and generating insight (El-Hadi & Merino, 2023). However, as we transition into a digital age, a pivotal question arises: What happens to these essential "deep reading" processes when confronted with digital-based reading modes? This is not a simple, binary debate of print versus digital reading but rather a complex issue that society must grapple with in this transition period.

The reading circuit is not hardwired into our genetic blueprint like vision or language; it requires a nurturing environment to develop and adapt. This environment shapes the circuit's development, favoring characteristics aligned with the dominant medium. In today's digital era, speed, multitasking, and processing vast amounts of information are prioritized. As a result, the reading circuit is subtly adjusting to these requirements, with less time allocated to slower, time-intensive deep reading processes like inference, critical analysis, and empathy. Evidence from educators and researchers in psychology and the humanities underscores these concerns. College students are increasingly avoiding classic literature because they find longer, denser texts challenging and demanding more patience. Yet, it's not merely cognitive impatience at play; there's a growing worry that many students lack the critical analysis skills needed to comprehend complex texts, whether in literature, science, or civic matters.

Studies also reveal that digital screen reading is associated with troubling effects on reading comprehension, beginning as early as fourth and fifth grades. Skimming and browsing are becoming the "new norm," with readers adopting an F or Z pattern, rapidly scanning text. This skimming behavior reduces the time spent on deep reading processes, making it difficult to grasp complexity, understand others" emotions, appreciate beauty, and generate original thoughts. Physicality is another crucial aspect of traditional reading. Print reading engages our sense of touch, providing a spatial and geometric dimension to words. This tactile element allows for recurrence, the ability to revisit and reassess one's understanding of a text, fostering comprehension. Further research indicates that screen reading can affect the growth of empathy in young readers, and these consequences extend beyond the realm of youth, impacting society as a whole. The erosion of critical analysis and empathy may lead to information silos, where individuals are susceptible to misinformation and demagoguery.

Nevertheless, there's hope on the horizon. The principle "use it or lose it" applies to the reading brain, implying that we have a choice in preserving its critical thinking capabilities. While the transformation of the reading brain is ongoing, science and technology offer the tools to identify and rectify these changes. By understanding what we stand to lose and embracing the new opportunities that digital mediums bring, we can cultivate a "bi-literate" reading brain. Such a brain will be adept at deep thought in both digital and traditional formats, ensuring that we can navigate the sea of information with wisdom and knowledge, fostering a thriving society. This endeavor is vital not only for us but also for the generations that will follow.

AI Literacy and Screen Literacy

The *Oxford English Dictionary* (*OED*) defines literacy as follows: "The quality, condition, or state of being literate; the ability to read and write." This traditional definition of literacy has been foundational for generations, emphasizing the fundamental skills of reading and writing. However, in our rapidly evolving digital age, the concept of literacy has undergone significant transformations, extending beyond traditional boundaries. Today, literacy encompasses not only the ability to read and write but also proficiency in various digital domains, notably AI and screen literacy.

AI literacy represents the capacity to understand, interact with, and effectively use AI technologies. It encompasses a range of knowledge and skills necessary for navigating the AI-driven world we live in today. Importantly, AI literacy transcends age, gender, ethnicity, and language conventions, making it accessible and relevant to individuals from diverse backgrounds. AI literacy has become essential in our time due to the increasing integration of AI technologies into our daily lives, ranging from virtual assistants like Siri and Alexa to recommendation algorithms on streaming platforms and the automation of various industries. An AI-literate individual can proficiently utilize AI-driven language translation tools, comprehend the underlying principles, and evaluate the quality of their translations.

AI literacy involves the following operational traits:

Understanding AI concepts: AI literacy quantitatively entails grasping core AI concepts, such as machine learning, neural networks, algorithms, and the mechanisms through which AI systems learn from data. A measure of this understanding could include the ability to explain these concepts and answer related questions accurately.

Using AI tools: Measuring AI literacy in this aspect involves evaluating an individual's practical skills in using AI-powered applications and tools for tasks like data analysis, automation, decision-making, and problem-solving. This could involve performance-based assessments or task completion metrics.

Addressing ethical considerations: AI literacy means recognizing and comprehending the ethical implications of AI. A methodological approach here could include surveys or questionnaires assessing an individual's awareness and understanding of AI ethics, along with their ability to identify ethical concerns in AI applications.

Cultivating critical thinking: To quantitatively evaluate critical thinking specific to AI, one could employ tests or assessments that require individuals to critically assess the credibility and reliability of AI-generated information and outputs. Scenarios or case studies involving AI could be used to measure this skill.

Alongside AI literacy, we find that screen literacy pertains to the ability to comprehend, navigate, and critically engage with diverse digital screens and interfaces, encompassing computers, smartphones, tablets, and other electronic devices. In our contemporary era, screen literacy is indispensable because screens have emerged as the primary medium through which we access information, communicate, work, and entertain ourselves. For example, a screen-literate individual can confidently navigate social media platforms, evaluate the credibility of online news articles, and implement measures to secure their digital accounts.

Screen literacy involves:

a) **Digital navigation:** The skill of effectively using and navigating digital interfaces, including software applications, websites, and social media platforms.
b) **Media literacy:** The capability to critically assess and analyze digital content, such as text, images, videos, and news articles, with considerations for factors like credibility and bias.
c) **Cybersecurity awareness:** An understanding of how to safeguard personal information and devices from online threats and cyberattacks.
d) **Communication skills:** Proficiency in communicating and expressing oneself clearly and appropriately within digital environments.

While the traditional definition of literacy emphasized reading and writing, our contemporary understanding of literacy has evolved to include AI and screen literacy skills, reflecting the evolving demands of our digitally driven society, even influencing the way we spell and communicate.

AI Native: Born Super-digital

First, some background: The term "digital native" was coined in 2001 by Marc Prensky. In his article *Digital Natives, Digital Immigrants*, Prensky (2001) defined "digital natives" as young people who grew up surrounded by and using computers, mobile phones, and other tools of the digital age. The devices and technologies that Prensky was referring to at that time are greatly different from the devices and technologies now. After all, in 2001, we only had dial-up internet connections and primitive computers and mobile phones.

Fast forward to today, where technology has dramatically evolved beyond basic mobile phones and dial-up connections, and the term "AI native," as explored in Kiaer (2023c), appears more apt. This refers to a generation that grew up in a seamlessly integrated digital world, where the lines between the physical and digital have blurred. AI natives are individuals who are born into a world where AI is an inherent part of daily life. Indeed, they are not only familiar with AI, but are inherently fluent in AI technology due to their lifelong exposure. AI natives are characterized by having been born into a world where AI technologies are ubiquitous and seamlessly integrated into everyday life. In addition, while AI natives may have an innate comfort with AI interactions,

AI literacy education can further enhance their understanding of AI concepts, ethics, and critical thinking specific to AI applications.

In essence, AI literacy and AI natives complement each other in our AI-driven world. AI literacy equips individuals, regardless of age or background, with the skills to navigate and understand AI technologies, while AI natives provide insights into a future where AI interaction is as familiar as breathing. Together, they shape our evolving digital landscape and the way we interact with AI.

A study encompassing 100,000 families from various regions, including the United States, the United Kingdom, and Spain, shed light on the online habits of children aged four to fifteen. The data, spanning from February 1, 2019, to December 31, 2020, dissected their preferences across five app categories: online video, social media, gaming, education, and communication. A standout observation was TikTok's rise in popularity, surpassing platforms like Instagram and even Facebook in some regions, illustrating the shifting online landscapes for young users.

Try to visualize children for whom touch screens, voice assistants, and virtual classrooms are as routine as handwritten letters were to previous generations. Their inherent tech-savviness often overshadows even those of older age groups. The digital world isn't an extension for them; it's an integral part of their existence. However, their deep immersion brings along challenges. Even as toddlers, they are handed devices, which, while serving as potential learning assets, can also become distractions. It's not uncommon for a modern child to be more familiar with a digital rendition of *Humpty Dumpty* than a bedtime story from their parents. The digital age, though bursting with educational prospects, presents dangers such as excessive screen time, exposure to unsuitable content, and decreased physical activity. The emphasis on digital safeguarding for children, especially those under the age of nine, has never been more crucial. Modern devices offer parental controls that restrict harmful content access and help manage screen time. As we guide children about real-world risks, we must also arm them against digital threats. This includes monitoring their digital activity, ensuring age-appropriate content exposure, and being vigilant for signs of cyberbullying. In this rapidly evolving digital age, our collective responsibility is to ensure that children harness technology's advantages while being protected from its potential harms.

Metaverse for AI Natives

The metaverse serves as a space not only for Gen Z, Alpha, and future generations but also for older generations who may find themselves compelled to embrace the new digital way of life. Still, it is the younger generations—Gen Z and Alpha—who are particularly attuned to and comfortable with the metaverse. Gen Z refers to individuals born between the late 1990s and the early 2010s. This generation follows the Millennial cohort (Gen Y) and precedes Gen Alpha. Growing up in a digital era with advanced technology and widespread internet access, Gen Z is known for being tech-savvy, socially conscious, and inclined toward digital communication and social media platforms. Their perspectives, values, and behaviors have been significantly influenced by rapid technological advancements, economic uncertainties, and social changes.

In a spring 2017 study, 70 percent of US children between the ages of eight and fifteen reported being "extremely" or "fairly" interested in experiencing VR, and 64 percent of parents reported the same (Aubrey et al., 2018). Gen Alpha comprises individuals born from the mid-2010s onward. This generation is still young, and its defining characteristics continue to evolve. It is anticipated that Gen Alpha will grow up in a world that is even more technologically advanced and globally connected than that of their predecessors. Being born into an era marked by the prevalence of artificial intelligence, virtual reality, and automation, Gen Alpha is expected to possess a high level of digital proficiency and a natural affinity for emerging technologies. The age at which children become digitally active is constantly decreasing. According to a study by Childwise, the majority of children now possess a mobile phone by the age of seven (PA Media, 2020). Other reports showed that more than 95 percent of the Gen Z population own a smartphone, 83 percent own a laptop, and 78 percent own an internet-connected gaming console (Howarth, 2023a). In comparison, older Millennials received their first cellphone at an average age of twenty, while younger ones received theirs around sixteen (Howarth, 2023a). In addition, a study by Childwise also found that many children now express fear of being without their phone, with over half admitting to sleeping with it by their bedside (PA Media, 2020). The study further revealed that children spend approximately 3 hours and 20 minutes each day engaging in activities such as messaging, playing games, and being online (PA Media, 2020).

My youngest daughter, Jessie, was born in 2012, and her schooling was interrupted by Covid-19. For almost two years, she took lessons virtually—that is to say, that one-third of her primary school education was online. Even now, a lot of her schoolwork requires using the internet. For her, it's the norm. From an early age, she has been able to use a computer and iPad, making do with emojis when she could not yet write. She and her friends play together on *Roblox*, a metaverse platform which most parents do not even know about. They were introduced to autocorrect mechanisms from a very young age. Jessie often asks me why she should learn spelling when Siri can always tell her what the correct spelling is. This isn't an easy question for me to answer. She, along with many of her friends, meet their overseas grandparents and relatives digitally more than physically. Jessie and her friends' lives are constantly accompanied with some form of AI, whether consciously or not. It's the way their lives are formed.

For AI natives like Jessie, AI is their playmate and VR is their playground; it's an integral part of their day-to-day experience, which of course involves learning. For them, simply banning AI or the usage of AI would seem largely unfair. It would be like banning the dictionary for the pen-and-paper generation. The issue is not whether to give them access to AI or not, whether they need to learn about AI is not even a question. Undoubtedly, we need to teach them and help them to use AI. Teachers should advise AI natives not to rely on AI entirely but to use it partially to maximize their performance. It is also the teachers' role to teach AI natives that AI, though named "intelligent," can lead them down the path of ignorance. To educate the younger generations, teachers need to understand the digital bliss and digital trouble that AI could engender.

Although Gen Alpha and future generations will live with AI in their daily lives, when we introduce them to incredibly capable and powerful AI like ChatGPT, we need to do so gradually. Such AI should be introduced to their lives later, if possible, except in some unusual circumstances. For these younger generations to be truly AI competent, they need to be acquainted with basic learning practices and their mind needs to grow and mature first to protect themselves from dangers like deepfakes and false information.

Gen Z and Alpha: Empowered or Endangered?

The metaverse provides an empowering but also endangering space. Its potential uses vary according to demographic. For young people, it offers opportunities in gaming and education. As one grows older, it becomes relevant for business and healthcare. Regardless of age, people are embracing this space as a means to relax and enjoy gaming, touring, and entertainment in virtual spaces. Nevertheless, one thing is clear—the physical world cannot be entirely replaced by the virtual world. Humans are physical beings, and our eyes have limitations in the virtual realm. Our basic physical needs must be met in a physical way. Simply put, we cannot live in the virtual world without food and sleep. In the following, the impact of the metaverse on our young generation will be explored.

Generation of Shorts: Understanding Super-Short Attention Spans

Research conducted by Microsoft in 2015 revealed that Gen Z individuals have an average attention span of only 8 seconds, 4 seconds less than that of Millennials. This decrease is attributed to the generation's continuous exposure to digital content and the prevalence of multiple screens. Our attention spans are intricately linked to the influence of social media and the role of dopamine. On average, a person's attention span extends to about 20 to 30 minutes. However, social media has profoundly impacted our collective attention spans. Information on these platforms is consumed rapidly, and trends come and go in the blink of an eye. What's popular on social media one moment can be easily forgotten shortly after.

According to neuropsychologist Dr. Sanam Hafeez, when we engage with platforms like TikTok, our brains actively seek dopamine, a neurotransmitter associated with pleasure and motivation (Thorpe, 2021). The encounter with content that evokes laughter or positive emotions triggers the release of dopamine, motivating us to seek more. This constant pursuit of dopamine often leads to frequent scrolling and switching between content. Dopamine also plays a crucial role in our attention. Research from 2016 in *Current Biology* indicates that when individuals experience a dopamine boost, they become more inclined to pay attention to similar stimuli in the future. Social media platforms such as Instagram, Snapchat, Twitter, and TikTok are designed to capture and retain our attention, albeit not necessarily for extended periods. While social media can affect short-term attention, its impact on long-term attention spans for adults versus adolescents may differ. Attention involves various brain regions, including those responsible for decision-making and rewards, and researchers are still exploring how social media influences these processes. Studies have highlighted that social media can foster impulsivity, with users often clicking away without second thought. Multitasking on social media, such as simultaneously using TikTok, Instagram, and Twitter, can erode our ability to filter out distractions and undermine our focus. Some research even suggests that people lose interest in content after just 8 seconds if it fails to engage them.

The fast-paced nature of social media and the pursuit of dopamine can significantly influence our attention spans. While it may impact short-term attention and impulsivity, the long-term effects on adult attention spans are an ongoing area of study. It is imperative for us to be mindful of our social media usage and its potential impact on our ability to focus and filter out distractions.

Zooming out to the bigger picture, we need to make a more concerted effort to create a balanced and harmonious system integrating the real world and digital realm, particularly for our younger generations.

VR: Reducing Anxiety and Learning to be Social

That said, VR can provide an empowering space. Research shows that VR has the potential to create spaces that foster empathy, fun, and imagination, with limitless applications. Indeed, the world's greatest museums will start having virtual twins (Rosen, 2021). Many educators are excited about the possibility of using VR to promote prosocial behavior among younger children, and 62 percent of parents believe that VR can enhance their children's educational experiences (Aubrey et al., 2018). Studies additionally suggest that VR can help older children develop perspective-taking skills, reduce racial bias, and cultivate empathy (Barbot & Kaufman, 2020). VR can reduce anxiety—particularly social anxiety related to speaking and interaction—making it a valuable tool for individuals returning to social situations. Gen Z spends a significant amount of time on social media, which can be a source of anxiety and distress but also support and solidarity. In a 2019 report, Hill Holliday, its research arm Origin, and media agency Trilia released an updated research study about Gen Z through its independent consumer and business insights research group. The study found that 48 percent of Gen Z individuals feel anxious, sad, or depressed because of social media, while 58 percent are "seeking relief" from social media (Hill Holliday et al., 2019).

An Ofcom survey found that only 15 percent of 16- to 24-year-olds prefer phone calls, while 36 percent favor instant messaging due to its quick and easy nature (Ofcom, 2016). Phone phobia is prevalent among young people, as they find it uncomfortable to interact with others in real-time. Phone phobia may reflect not only their fear of interacting with people in real life but also some other social aspect of Gen Z. As Forbes writer Brianna Wiest explains, "Phone calls seem invasive because they demand an instant response" (Wiest, 2019). There's a sense of entitlement in a phone call, a demand for the caller to acquiesce to your timeline. It could be that Gen Z and Millennials are just more aware of manners and politeness. A VR space that bridges the gap between real life and traditional online platforms can be a good solution to this. Introverted students may not find it easy to flourish in the classroom, but they seem to feel much more comfortable and perform better in the VR classroom, where they can present themselves as avatars of their choice.

VR: Time Limits

The impact of VR depends on the platforms and how they are used. It is important to note that excessive use of VR can have negative effects. Following suggestions from Jeremy Bailenson, the head of Stanford University's Virtual Human Interaction Lab, VR sessions should be limited to 20 minutes, with regular 3-minute breaks, hydration, exposure to natural light, and reminders of the physical body's presence.[1] For children, moderation is important. Instead of hours of use, which might apply to other screens, think in terms of minutes. Most VR experiences are meant to be limited to the 5- to 10-minute scale (Aubrey et al., 2018). When it comes to content, a good rule to follow is that if you wouldn't want your children to live with the memory of the event in the real world, then don't have them engage in it in VR. While activities like traveling to the moon are fine, exposing them to scary experiences may have a lasting impact. Additionally, it is important to consider practical safety. VR, by definition, blocks out the real world, so it's crucial to monitor your children around physical hazards such as sharp edges, pets, and walls.

VR for Young Children?

Researchers have warned that VR headsets could pose risks to users, particularly children (University of Leeds, 2017). Scientists based at Leeds University believe that continued use of VR sets could trigger eyesight and balance problems in young people unless changes are made to devices. Leeds University professor Mark Mon-Williams states, "In a VR device, a virtual three-dimensional world is displayed on a 2D screen, and that places strain on the human visual system. In adults, that can lead to headaches and sore eyes. But with children, the long-term consequences are simply unknown" (Arguinbaev, 2017). Immersive virtual reality also interferes with default head-trunk coordination strategies in young children (Miehlbradt et al., 2021). In their guidelines, the World Health Organization (WHO) recommends that children under the age of one should have no screen time, while children between the ages of two and four should have no more than 1 hour of sedentary screen time (World Health Organization, 2022). These recommendations are part of a broader set of guidelines on physical activity, sedentary behavior,

---

[1] Adapted from Eight Rules to Help You Stay Safe in Virtual Reality (Bailenson, 2018).

and sleep for children under five years of age. Many VR device manufacturers often recommend that their products not be used by children under the age of twelve or thirteen due to potential health concerns, such as vision problems or cognitive and developmental issues (Linkedin, n.d.). Given the immersive nature of VR, it's crucial to monitor usage closely, ensuring that the device is used safely, and that usage is balanced with physical activity and other offline activities. As always, parents and caregivers should supervise children when using these types of devices.

Children's increasing interaction with AI-driven metaverses raises profound questions concerning social, cognitive, and language development. The potential of AI in the metaverse to replace core human attributes poses a significant quandary. Typically, children acquire language skills through interactions with fellow humans within a supportive and nurturing environment. Language acquisition is widely acknowledged as being reliant on social engagement, emotional connections, and exposure to meaningful contexts. It seems unlikely that children can solely rely on interactions with generative AI, like ChatGPT, in VR settings for language acquisition. However, the metaverse introduces a captivating realm for exploration.

While children may not acquire language and other social and cognitive behaviors in the metaverse as they do through real-life interactions, VR does offer avenues for language exposure, practice, and cultural immersion. Within the metaverse, young individuals can actively engage with virtual communities, participate in diverse activities, and communicate with others across an array of virtual environments. This opens the possibility of nurturing awareness of diversity. The immersive and interactive nature of the metaverse empowers children to engage with various languages, cultures, and perspectives, thereby contributing to their overall language awareness and comprehension. Although the metaverse may not perfectly replicate the richness of linguistic environments found in real-life interactions, it can serve as a complement to language learning experiences, facilitating language practice and exposure.

Yet it almost goes without saying that young children could become confused if they don't learn how to engage with real people and live in the real world, leading to delusion. Too much time spent in online and virtual worlds may result in a lack of knowledge on how to communicate and gesture effectively. To the extent that VR simulates the real world, children may face challenges discerning which components of virtual events are not real (such as seeing the self in VR versus seeing another person in VR) (Segovia & Bailenson, 2009). Preliminary findings indicate that the illusion of VR has a greater impact on young children in comparison to adults. Furthermore, the effects of VR tend to be magnified when compared to traditional media, such as TV. An experience in VR, where individuals are perceptually surrounded and interact with the scene using natural body movements, tends to have a more significant impact than a similar experience using other media.

The effects of the use of immersive VR on young children's still-developing brains and health are unknown, though most parents are concerned, and experts advocate moderation and supervision. Researchers say that there is still much to learn about how VR impacts children's brains. In a 2008 study, Baumgartner et al. found that adults experiencing virtual reality were able to use their prefrontal cortex to regulate their brain processing. In contrast, the children were not able to do so, and thus, they couldn't always distinguish between what was happening in a virtual world and what would happen in real life. As Thomas Baumgartner, a neuroscientist at the University of Zurich and author of the study, explains, "their prefrontal cortex is far from being fully developed at this age" (Trageser, 2022). As virtual worlds become even more immersive and realistic, the line between virtual and reality will be even harder for children to discern. Spending too much time in a virtual environment could lead children to form memories based on virtual experiences, which may result in unrealistic expectations of reality, such as how their bodies or houses should look, and a potential loss of real-world social skills. More studies are required to understand the full impact. Not only are there physical and mental problems associated with VR, but there are also ethical issues that I will address in Chapter 3. One clear message is to wait and slow down, if possible. However, for educators and researchers, it is crucial to thoroughly investigate the impact of virtual reality as soon as possible. There is no time to lose.

Too Playful?

In my previous book, *Emoji Speak* (2023), I discussed how our communication is changing from text-based forms to image-based, multimodal, and multisensory forms. We have lived in a black-and-white space with text alone for too long. Technology is liberating us and opening new avenues of communication that closely depict our face-to-face, real-time interactions. Simple face emojis can be playful, and various memes add a comedic touch to our communication. They have the ability to lower communication barriers between individuals, whether they are familiar or unfamiliar to each other. Nonetheless, some teachers express concerns that the cute and often comic nature of emojis, memes, avatars, and the metaverse can create an impression of insincerity. Fundamentally, they worry that digital spaces suggest play rather than learning. While gamification is an effective and enjoyable way for us to learn, it may also contribute to the perception that people are less serious in the metaverse. Personally, I find it hard to imagine a criminal court in the metaverse conducted with avatars.

AI Intervention

The last attribute of the super-digital era is generative AI. In the era of advanced digital technology, the presence of AI—particularly interactive and generative AI—has become accessible to the public. However, the role of AI is unclear. Is it a helper, a dictator, or perhaps an enemy? Recent advancements in technologies like ChatGPT have made the questions we once encountered in science fiction novels feel all too real. Nowadays, I often experience moments where, while having conversations with people in the "real" world, my Apple Watch or iPhone unexpectedly intervenes or interrupts, mistakenly activated by my words. Initially, it was merely annoying, but as it continued, I started feeling increasingly uncomfortable. Is AI listening to my conversations so closely?

The metaverse, an expansive digital universe, is on the cusp of a transformative shift with the entrance of ChatGPT. Devised by OpenAI and launched in November 2022, ChatGPT is a product of the GPT-4 model. Presently, ChatGPT is experienced in a text-based format. However, its underlying architecture and potential go far beyond simple textual exchanges. OpenAI's decision to make its core foundation open-source allows developers globally to harness and expand its capabilities. With this, it won't be long before ChatGPT could possibly communicate not only through text but could also possess a natural-sounding voice or even a visual representation in the metaverse. Such rapid advancements bring us closer to the concept of the "singularity"—a theoretical point where technological growth becomes uncontrollable and irreversible, leading to unforeseeable changes to human civilization (de Weck, 2022, p. 613). The concept of AI singularity, where artificial general intelligence (AGI) surpasses human intellect, has traditionally been seen as decades away. However, recent insights from AI researchers, including voices from firms like Anthropic and OpenAI, indicate we might be on a faster trajectory toward this moment. Feedback loops in AI research and exponential growth patterns suggest the transition to AI dominance may be imminent. Notably, even the creators of ChatGPT believe that future AI tools could make current systems look rudimentary. While the potential of AGI-level systems is immense, so are the challenges ahead. The line between real-human and digital-human interaction becomes increasingly blurred, challenging our traditional notions of identity, consciousness, and interaction. In the evolving educational landscape, a digital human version of ChatGPT might serve as an omnipresent instructor, while in administrative tasks, it could function as a diligent digital secretary. Yet it's not just about its adaptability; the arrival of ChatGPT signifies an acceleration into the super-digital era, a time where the digital and the physical worlds are more intertwined than ever. However, with great power comes great responsibility. The potential misuse, such as the spread of fake news or misinformation, raises serious ethical questions.

There is a growing belief that we are on the verge of achieving human-level artificial intelligence. Dr. Nando de Freitas, a lead researcher at Google's DeepMind AI division, stated that the decades-long quest for artificial general intelligence (AGI) is nearing its end after DeepMind unveiled an AI system capable of complex tasks, such as stacking blocks and writing poetry (Cuthbertson, 2022). Described as a "generalist agent," DeepMind's new Gato AI only requires scaling up to create an AI that can rival human intelligence, according to Dr. de Freitas (Cuthbertson, 2022).

The integration of digital technologies is rapidly becoming the norm in all aspects of our daily lives, expanding our digital footprint. With rapid advancements in hardware and software, as well as the emergence of new applications and AI technologies, our lives are being transformed. However, the development of generative AI, exemplified by ChatGPT, has raised concerns about these digital tools transitioning from being mere assistants to potentially monitoring and dictating our actions.

The fear of AI replacing human labor is not unfounded. Data from Challenger et al. (2023) reveals that AI was responsible for approximately 4,000 job losses in the previous month alone. The interest in AI's potential to perform complex organizational tasks and reduce human workloads continues to grow. According to a report by the outplacement firm, layoff announcements from US-based employers in May 2023 exceeded 80,000, representing a 20 percent increase from the previous month and nearly quadrupling the figure from the same month the previous year (Challenger et al., 2023). Among these job cuts, around 3,900—or approximately 5 percent—of total job losses were attributed to AI, making it the seventh most prevalent factor contributing to job loss in May, as reported by employers (Challenger et al., 2023).

It is evident that we can never go back to the pre-ChatGPT era, as AI has become an integral part of our lives. The crucial question that remains to be answered is whether we will be able to maintain our autonomy and control in this new digital landscape. As we navigate this evolving digital world, it is imperative to address the ethical, legal, and societal implications of AI's increasing influence to ensure a future that respects human values and safeguards our autonomy.

The AI Genie: Helpful or Harmful

AI has become incredibly popular, so much so that it was named the word of the year by Collins Dictionary (BBC News, 2023a), and its prominence is only destined to increase. Young children, whether permitted or not, are incorporating AI into their activities (BBC News, 2023b). The AI symbol, signified by the spark icon, begins to appear as you work on your documents, almost akin to Aladdin's magical lamp. It's as if a genie emerges from it whenever you require assistance or guidance in your tasks. The question of whether this is good or bad is not a simple one.

Digital Injustice

Recent data (Kemp, 2023) illustrates the disparities in active social media use across different regions of the world. In Western Africa, the percentage of active social media users is 14.5 percent, while in Northern America, it is significantly higher at 73.7 percent. It is important to note that the lower percentage in Western Africa does not necessarily reflect a lack of interest in social media but rather highlights the challenges related to access to digital devices and internet connections. This digital divide is likely to become more pronounced, particularly when considering the scarcity of VR headsets and other wearable devices in many parts of the world.

Addressing digital divides and making technology more accessible globally is crucial to prevent digital injustice. The metaverse holds the potential to be an enabling space, capable of transcending physical prejudices. However, it is essential to ensure that the metaverse becomes affordable and accessible to people from all walks of life. Otherwise, there is a risk of creating another layer of injustice and prejudice, where certain individuals or communities are excluded from the benefits and opportunities offered by the metaverse. Efforts should be made to bridge the affordability gap and provide equal access to technology and digital platforms worldwide. This will help foster inclusivity, empower marginalized communities, and ensure that the metaverse becomes a space where everyone can participate and thrive, regardless of their socioeconomic background or geographical location.

From Nation-States to Tech Empires: The New Landscape of Digital Governance

In today's increasingly digital landscape, social media platforms are rapidly evolving beyond their initial roles as communication tools or content-sharing spaces. With an eye on monetization, many of these platforms are now aspiring to become all-in-one packages, offering services that range from e-commerce to financial services like banking. On the one hand, this trend embodies the spirit of transnationalism, knitting together a global culture that transcends national borders. Yet, on the other hand, it poses critical questions about governance and the potential biases of digital capitalism. Social media platforms are ceaselessly expanding their range of services. What started as spaces to connect with friends and share photos are now morphing into marketplaces, news outlets, and even financial service providers. Companies like Facebook, for instance, are not just social networks anymore; they are venturing into the realms of digital payments, online stores, and more. This expansion is not merely a response to consumer needs but a calculated strategy for monetization. By offering an all-in-one package, these platforms aim to make themselves indispensable to users, thereby increasing their revenue streams and market influence.

Simultaneously, the digital age has also given rise to a new kind of transnationalism. Social media platforms, with their global reach, have enabled the creation of communities that are not bound by geographic locations or national identities. This new form of transnationalism encourages the sharing of ideas, culture, and information on a global scale, creating a diverse and inclusive virtual world. It's a notion that is seemingly beneficial, helping us build a culture beyond nations and forging global unity through digital connections. However, this amalgamation of features and global reach comes with its own set of challenges, particularly in governance. When social media platforms offer services like banking, they start resembling not just tech companies but financial institutions. Yet, unlike traditional banks, these platforms are not necessarily subject to the same level of regulatory oversight. This raises questions about accountability, data privacy, and financial security. Moreover, this expansionist trend is also setting the stage for what can be termed as "digital capitalism," where the influence of these platforms extends beyond just the social or technological realms into the economic and political. It can essentially make capitalism the superpower, guided by the biases and objectives of big tech companies. These platforms can, in theory, dictate economic terms and conditions that impact millions, all while potentially circumventing local governance structures.

Digital Capitalism and Affordability

VR, while revolutionary, requires sophisticated hardware, most notably VR headsets. The data indicates that by 2022, 37.7 million Americans owned such devices (Cureton, 2023b). When translated to a global scale, the numbers portray a stark divide: While advanced nations like the United States have a significant proportion of their population entering the metaverse, what about developing or less affluent nations? The high cost of VR headsets means they remain inaccessible for many, creating a new form of inequality—those who can afford to be part of the digital renaissance and those who remain on the outskirts. In essence, while the metaverse has the capability to transcend geographical barriers, it could inadvertently establish economic ones. Currently, the burgeoning metaverse ecosystem might lure users with the promise of free access. Yet, as history with various digital platforms shows, sustaining a free model can be challenging. The shift to monetized models, such as subscription fees or premium content behind paywalls, is often inevitable. This raises serious concerns about the democratization of the metaverse experience. If pivotal aspects of the metaverse start requiring payment, it could exclude a significant portion of potential users, particularly those from socioeconomically disadvantaged backgrounds.

The essence of capitalism is the pursuit of profit, and as the digital realm grows, so does the scope for capitalistic ventures. The metaverse, if not carefully navigated, could become a playground for the affluent, where the rich not only have better access but also a superior, enhanced experience. This would inevitably lead to a two-tiered metaverse: one for those who can afford a premium experience and another, diluted version for those who can't. What is more, the dominance of certain tech giants could mean monopolization of the metaverse, leading to limited competition and further driving up costs. The presented data highlights the incredible potential of the metaverse and VR technology, but it also nudges us to reflect on the socioeconomic implications of such advancements. It's crucial for stakeholders—be they tech developers, policymakers, or consumers—to be aware of the risks of digital capitalism and strive to make the metaverse an inclusive space, free from the digital divides that plague other areas of the internet.

Remaining Challenges

Legal challenges, intellectual property, data security, and privacy all need to be resolved for the metaverse to truly resemble a real place. This task is far from easy, as evidenced by non-fungible tokens (NFTs). NFTs have emerged as unique digital assets stored securely on blockchain technology, completely transforming the notion of virtual ownership. Within the metaverse, individuals have the ability to own and trade NFTs representing a wide array of digital items, including artwork, virtual properties, fashion items, and even virtual identities. The appeal of NFT ownership lies in the exclusivity and verifiability of these tokens, as each one possesses distinct characteristics and can be traced on the blockchain. By owning an NFT within the metaverse, individuals can showcase their digital assets, express their personal style, and actively participate in the virtual economy. However, the rise of NFT ownership also presents complex questions and challenges. The convergence of the real world and the virtual world gives rise to legal implications, encompassing issues such as copyright, intellectual property, and jurisdiction. It is an ongoing endeavor to establish appropriate regulations and frameworks that govern NFT ownership and address conflicts between the real and virtual realms. The ultimate goal is to create a metaverse that truly feels like a real place, where humanity can thrive and engage in a secure and regulated environment.

3

Metaverse Diversity

Digital Space Shift: Screen, Mobile, Then Metaverse

The digital landscape has experienced significant shifts over the years. Initially, it was oriented around large screens such as TVs and computers. However, with technological advancements, the prominence of mobile screens, including smartphones and tablets, has reshaped our digital interactions. Platforms like TikTok highlight this transition, as mobile-centric content and short-form videos have become a staple in social media and entertainment. Now, another shift is on the horizon with the emergence of the metaverse. Beyond the experiences of large screens and the mobile content of smartphones, the metaverse introduces a move toward highly immersive virtual environments. This realm, merging augmented reality, virtual reality, and other advanced technologies, offers multidimensional experiences tailored to individual preferences. Within the metaverse, the constraints of physical space are challenged, opening avenues for exploration, connection, and more. The progression from large screens to mobile devices, and now toward the expansive possibilities of the metaverse, underscores our digital evolution. As we approach this new phase, the distinction between the physical and virtual worlds might become even more nuanced, presenting a myriad of meanings and possibilities.

Web 3.0 and Diversity

The newfound diversity in the metaverse can be largely attributed to the emergence of Web 3.0 and its symbiotic relationship with technological advancements. Web 3.0, with its decentralized nature, paves the way for a more inclusive and diverse platform, empowering a broader spectrum of creators and developers (ACODS UK, 2023). Gone are the days when a handful of centralized entities set the parameters of the digital universe. Now, thanks to Web 3.0, the creation and ownership of the metaverse are more distributed, democratized, and diverse. Building on this foundational shift, new metaverse platforms have arisen. These aren't mere adaptations of their predecessors; they represent a seismic shift in the conceptualization, creation, and experience of virtual worlds. Further technological progress, encompassing advanced virtual reality, augmented reality, and AI-enhanced interactions, has added even more depth and realism to these platforms. Consequently, today's metaverse is a mosaic of distinct worlds, each offering unique experiences, ranging from hyperrealistic simulations to avant-garde, abstract dimensions.

In recent times, Google Search has undergone a significant transformation, providing users with multiple search modalities, such as a microphone and camera, instead of the traditional typing method as seen in Figure 3.1. This shift reflects the evolving way we interact with languages, embracing a more multimodal approach. With the introduction of voice commands, image recognition, and visual searches, the way we interact with technology and languages is changing rapidly. The convenience and efficiency of using microphones and cameras to convey our queries and search for information have made typing seem somewhat antiquated in certain contexts. This trend in multimodal language usage is reshaping how we communicate with technology and each other. As we adapt to these new methods, we are witnessing a significant shift in how information is accessed and processed. The integration of various modes of communication is becoming the norm, reflecting a more dynamic and inclusive approach to language and technology.

We can draw a parallel with fax machines. In the not-so-distant past, fax machines were an essential tool for transmitting documents across long distances. However, as technology advanced and electronic communication

Figure 3.1 Images in the Google Search bar (Courtesy of Sarah Kiaer).

methods became prevalent, they gradually lost their prominence and are now rarely utilized, existing more as a relic of the past than a primary means of communication. This ongoing transformation in language usage and technological interfaces emphasizes the dynamic nature of our ever-evolving digital landscape. It also highlights the importance of embracing change and staying open to new communication modalities to keep pace with the constantly shifting technological landscape. As we move forward, the fusion of diverse communication methods will continue to shape the way we use languages, driving us toward a more integrated and interconnected future.

Idiolects in the Metaverse: A New Babel

In the realm of human communication, everyone manifests a unique linguistic blueprint, termed as an idiolect (Coulthard, 2004). This individualized pattern emerges from a combination of vocabulary, grammar, and pronunciation nuances (Whitehead, 2010). It's a tapestry shaped by various influences, ranging from one's geographical location to their socioeconomic background. Penelope Eckert's research has tracked noteworthy shifts in the domain of linguistic studies. She brought to the forefront the "Third Wave" of variation studies, which pivots its attention to the social implications of linguistic variables (Eckert, 2012). This approach weaves language styles closely with identity tags, offering a detailed look into how specific variables mold these styles. At the heart of this wave lies a pronounced focus on the inherent social connotations within language and its expressive components.

As we venture into the age of the metaverse, a realm where the digital coalesces with the tangible, the importance and complexity of idiolects and sociolects come to the fore. In this digital landscape, every utterance, every typed word, takes on new layers of meaning. This metaverse idiolect often transcends the boundaries of traditional languages, merging verbal expressions with symbols, emojis, memes, and other digital idioms. Such linguistic landscapes are frequently set by community members, creating a rich tapestry of expressive freedom. Yet, with this newfound linguistic liberation comes challenges. The multifaceted and eclectic nature of these new idiolects and sociolects not only enhances individual and group expressivity but also paves the way for potential confusion. Words and symbols, in this context, can be employed in an unconventional manner, giving them meanings that might be unfamiliar or even obscure to outsiders. This dynamism, while fostering creativity, might inadvertently become exclusive.

Those outside specific metaverse communities could find it challenging to decode and understand the nuanced expressions within.

The Third Wave, sensing this paradigm shift, expands its horizons to incorporate all linguistic expressions, regardless of their originating medium. One salient aspect of the Third Wave is its investigation into the nexus between speaker categories and their chosen linguistic personas (Eckert, 2012). While traditional studies might box individuals into predefined dialects or linguistic communities, the Third Wave takes on a more fluid stance (Eckert, 2012). Especially within the ever-evolving dynamics of the metaverse, this acknowledges that individuals, or even groups, frequently recalibrate their language patterns, tailoring them to specific contexts or audiences. Indeed, the dawn of the metaverse has, in many ways, revolutionized the realm of human languages. While, on one hand, it serves as a melting pot for languages and dialects, on the other, it accentuates the significance of idiolects, emphasizing that each individual carries their linguistic identity. Furthermore, the sociolect phenomenon—where group practices shape linguistic preferences (Lewandowski, 2010)—becomes even more pronounced. This evolution is reminiscent of a modern-day Babel; while the metaverse offers a shared space for communication, it simultaneously amplifies our linguistic diversities, making the tapestry of human expression richer and infinitely more complex.

"Digilect": A New Language?

As discussed, new technologies have added to our communicative repertoire, allowing us access to novel modes of interaction that go beyond audio or text, like visual resources (such as emojis and images of the interlocutor on video calls) as well as a combination of various types of resources. For example, video calls can involve verbal and/or nonverbal communication. GIFs can move and have text on them, thus combining the visual, the textual, and sometimes the auditory too. This has given rise to what Veszelszki (2017) has termed a "digilect"—a type of virtual language with its own features that can affect other forms of language, such as oral and handwritten communication. It has even been suggested that in a century from now, our lives may become so deeply intertwined with the virtual world that it becomes perceived as identical to the physical world. This could mean that "technologically mediated" communication would be comparable to and even supplant "non-technologically mediated" communication (Osler & Zahavi, 2022, p. 2).

While there are no concrete statistics for how much of human communication is dominated by virtual interactions on a global scale, the amount of internet usage could be a good indicator of how much human interaction is mediated by technology. A study on internet usage in 2022 found that the average internet user spends around 7 hours a day online, or approximately 40 percent of their waking hours, and this number is projected to continue increasing (Kemp, 2022). It is also estimated that some 5.18 billion and 4.8 billion people are internet and social media users, respectively. This accounts for 64.6 percent and 59.9 percent of the global population, respectively, which both make up majorities. Thus, it is probably not an exaggeration to say that most people in the world spend a considerable proportion of their day communicating online in some form. Brown Sr. (2020, p. 232) warns of the danger of online communicative technologies, arguing that technology usage is "almost out of control," garnering the "undivided attention" of its users to the extent that face-to-face communication is no longer the main mode of interaction.

As I sat in the digital humanities workshop, admiring the sketches drawn by my students, one piece caught my attention. The drawing in Figure 3.2 encapsulated a sentiment that echoes throughout the digital sphere: the shock, awe, and intimidation incited by the breakneck speed of AI advancements.[1] It was as if this student had managed to visually represent the thoughts swirling in my head—that the metaverse woke me up. A new world has been born, and we are struggling to keep up. Recent remarks from the UK's security minister, Tom Tugendhat, resonate with this sentiment. He warned that the pace of AI is so swift that regulatory mechanisms can't keep up (Gibbons, 2023). A shift in our cultural and regulatory approach is crucial to address this emerging challenge. It isn't just him either; industry pioneers like Elon Musk and Steve Wozniak have shared their apprehensions (Kahn, 2023). A testament to this widespread concern is the inception of movements like *PauseAI*. The core message behind such movements is clear: While the evolution of AI is admirable, it's time to consider a slight deceleration. The overarching fear? Once this technological train departs at full speed, halting or even slowing it down might be an insurmountable challenge. A significant portion of the US public, with numbers oscillating between 64 percent and 69 percent, side with this cautious approach (PauseAI, n.d.). Their apprehension isn't unfounded; even among the intellectual elite, there is a consensus to reconsider our pace. The United States, especially

---

[1] Thanks to Szymon Uszok for granting me permission to use this image.

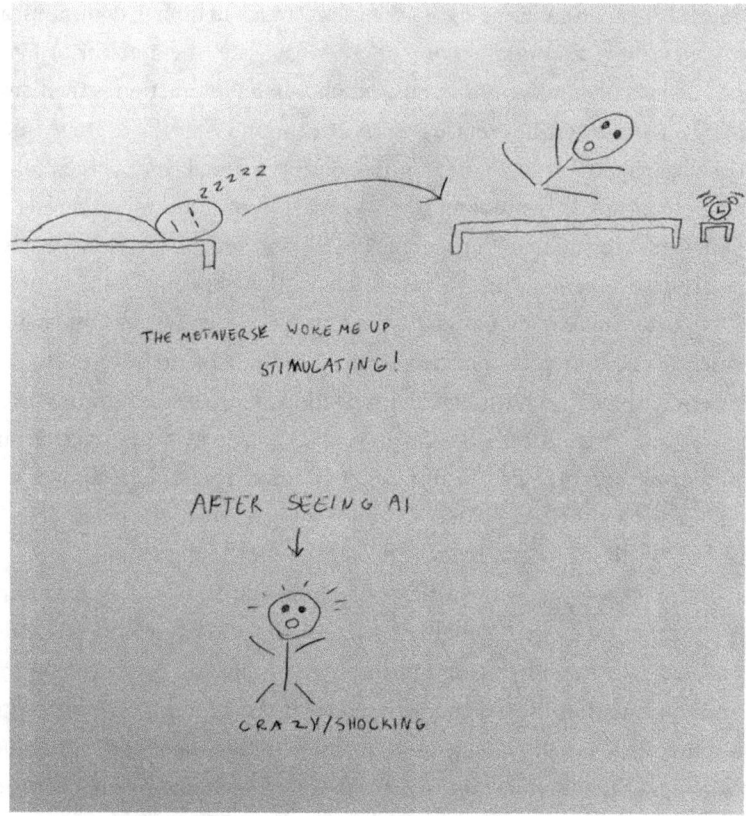

Figure 3.2 "The metaverse woke me up, stimulating!"

tech hotspots like California, given their pioneering status in the AI domain, is looked upon as a potential torchbearer for initiating a pause.

The warp speed of AI isn't just a domestic concern; its implications are global. This was the core spirit of the UK's AI Safety Summit in 2023. The summit emphasized an international treaty addressing AI's exponential growth, hoping for an endorsement from every UN member state. The establishment of an international AI safety agency, somewhat on the lines of agencies like the International Atomic Energy Agency (IAEA), has been mooted (Nichols, 2023). The agency would be responsible for the critical tasks of vetting AI deployments, training methodologies, and consistently updating the global community on safety protocols and research. In this rapidly evolving landscape, one thing is evident: Unchecked training of potent AI models without a consensus on safety is a path we should tread cautiously. AI's deployment should ideally happen under the keen watch of democratic mechanisms. This raises a fundamental question: Do we ride this tumultuous wave with reckless abandon, or do we step back, recalibrate, and then proceed with both excitement and caution? The ball, it seems, is in our court.

Robots Among Us: Navigating the Digital and Physical Convergence

The metaverse often captures our imagination, luring us into its intricacies, but the digital domain isn't the only place where AI and technology intermingle with our lives. There's a tangible fusion happening right in our physical surroundings, marked by robots taking on roles traditionally held by humans. Figure 3.3 depicts a robot designed to assist travelers at Incheon Airport in Seoul. At first glance, it's an ensemble of metal, circuits, and screens. Yet, its actions tell a more complex story. As it navigates the hustle and bustle of the airport terminals, it

Figure 3.3 Robot helped in Incheon Airport.

doesn't just provide assistance. It tries to communicate, attempting to strike a chord of familiarity by mimicking human-like behaviors. It might approach a lost traveler, its screen flashing facial expressions, perhaps a raised digital eyebrow of curiosity or a smile meant to comfort. Despite its earnest attempts, the robot's gestures remain distinctly mechanical, a stark reminder of the chasm between organic human interactions and programmed responses. It strives to be a part of human communication, yet its endeavors highlight a poignant reality: the challenge technology faces in truly emulating human nuances.

These robots, and many others like them, represent a broader trend. They're not just limited to the screen or virtual reality; they are very much a part of our physical world, nudging us to recognize the duality of our existence in an age of rapid technological advancement. In this intertwining of the digital and physical, moderation is crucial. While these robots, much like elements of the metaverse, offer convenience and innovation, our engagement with them should be tempered with awareness. Being discerning about where and how we allow technology to intervene in our lives ensures that we retain the essence of human connection, even as we stride forward into a technologically augmented future.

Diversity in the Metaverse: More Than the Inhabitants

The metaverse is an expansive and intricately woven realm that seamlessly melds our digital and physical existences. This confluence of realities is marked by a dazzling spectrum of digital diversity, allowing both humans and nonhumans to cohabit. But the metaverse's diversity extends beyond just the entities within it. It encompasses myriad forms of digital expressions—from avatars that stand as digital proxies for us to the array of emojis and memes used as conduits for emotions and ideas. Within this vast digital realm, diversity is showcased not only in the "who" but also in the "how." While entities or agencies in the metaverse—be they individual users with their avatars or sophisticated AI-driven systems—offer a plethora of perspectives and experiences, the methods of communication within these spaces have also undergone a remarkable transformation. In this digital age, the lines demarcating our physical selves from our digital personas are becoming increasingly indistinct. Individuals can now transcend their physical confines, crafting avatars that can mirror their real-world image or delve into the realms of fantasy. These digital representations, brimming with agency, are directed by human intentions or AI capabilities, enabling interactions that rival or even surpass their real-world counterparts. The means of communication have also evolved exponentially. Traditional modes, like letters or basic digital messages, now appear antiquated. The metaverse fosters immersive communication, where users engage in multimodal dialogues, blending text, voice, visuals, and even emotions. This fusion of communication modes not only enriches interactions but also mirrors the essence of the metaverse itself—a diverse, holistic, and evolving digital ecosystem.

Exploring Diversity in Metaverse Communication

There are various ways of communicating within the metaverse, each method offering a unique way to connect. Communication can take on many forms, from the traditional "one-person speaks while the rest listen" to more collaborative and dynamic interactions, where all participants express their opinions. Conversations can be one-on-one or group talks, with various formats accommodating the needs and preferences of the participants. One of the necessary choices to be made in metaverse communication is whether to present oneself authentically or to adopt an avatar representing an augmented version of oneself. This choice provides a way to customize and control how one appears to others within the virtual space. Sometimes communication happens with real names, while at other times, individuals might use nicknames or even communicate through avatars. The choice to use video by turning on a camera, to embody an avatar, or to appear as one's real self adds layers of personalization and anonymity to the experience.

Timing in the metaverse also offers flexibility; conversations can happen in real-time, or messages can be left for others to respond to at their convenience. These interactions can be enriched with multimedia resources such as videos, photos, emojis, and more. In recent decades, the adoption of communication technologies like instant messaging, video conferencing, live streams, social media, and extended reality has radically transformed the way in which we interact with one another. One of the most prominent changes is that these online technologies have extended the temporal and spatial possibilities of remote (i.e., non-face-to-face) communication beyond that of just real-time audio chats and nonsynchronous written communication. Now we can communicate anywhere at any time: Auditory communication does not have to be synchronous as we can leave audio messages that can be heard later on, textual communication through various instant messaging platforms can also be conducted in real time or asynchronously (if we choose to reply to messages later than they were received), and video conferencing technologies enable a combination of real-time visual and verbal communication.

While this diversity in communication methods broadens the scope of connection and allows for creativity and expression, it also poses challenges and risks. The freedom to present oneself in various ways can lead to identity conflicts or fake identities, and even criminal activities such as fraud. The anonymity that the metaverse can offer might sometimes provide a cover for malicious behavior. These complexities emphasize the importance of responsible engagement and a consideration of ethical practices within this expansive virtual environment. The metaverse offers a sense of community that can be tailored to individual needs and contexts, but it also requires vigilance to ensure that this exciting frontier remains a positive and safe space for all its inhabitants.

Beyond Words: The Journey from Alphabet to Multimodality

Victor Hugo once mused, "All characters were originally signs, and all signs were once images" and "Human society, the world, man in his entirety is in the alphabet" (de Looze, 2016, p. 132). This statement illuminates the central role of written words in our history, acting as the foundation of our understanding, influencing cultures, and forwarding progress. With the advent of technology and the surge of virtual immersion, our once predominantly text-bound communication has been liberated. We are no longer confined to the written script; our interactions have blossomed into a multimodal dance of audios, videos, GIFs, animations, and an array of multimedia components. This transformation resonates with our inherent desire for diverse and profound methods of expression, echoing our longstanding relationship with both spoken and unspoken communication. In my previous book, *Emoji Speak* (2023a), I delved into the potential intersection between visual and textual languages, suggesting a future where textual content might seamlessly integrate into the broader realm of visual communication, driven by the allure of visuals.

While Hugo highlighted the importance of the alphabet in human narratives, he also demonstrated an understanding of succinct, nontextual communication. A well-known anecdote narrates his terse correspondence with his publisher, Hurst and Blackett, in 1862 (LaFrance, 2016). Eager to know about the sales of *Les Misérables* while on vacation, Hugo penned a singular query: "?" The publisher's reply was equally brief: "!" The essence of communication persists, but with the progression of technology and our virtual migration, the mediums and tools have diversified, paving the way for a richer tapestry of expression. As the metaverse gains traction, the value of multimodal communication becomes indispensable. Historically, nonverbal cues like a reassuring smile or direct eye contact played crucial roles in face-to-face interactions. In the metaverse, these cues undergo a digital renaissance. For instance, the backdrop of video calls offers insights into our personal realms. Textual dialogues are enhanced by emojis, GIFs, and avatars, translating our tangible emotions and gestures into the virtual domain. Navigating the metaverse means assimilating these new communication norms. Something as natural as maintaining eye contact now translates to focusing attentively on a camera. Misreading or neglecting these digital nuances could lead to misunderstandings, whereas embracing the virtual world through expressive avatars or emojis can signify genuine engagement. In this realm, images, visuals, and gestures often rival the prominence of words. Proficiency in this evolving language, as we further immerse ourselves in the metaverse, becomes paramount. The varied communicative avenues of the metaverse enrich our discourse, emphasizing that sometimes, words in isolation fall short.

Moderation Matters: Navigating the Diverse Ecology of the Metaverse

Navigating the digital world offers undeniable convenience. However, it is crucial to approach it with moderation, grounding ourselves in principles that have guided human behavior for centuries. One such principle, deeply rooted in Confucian thought, is *zhongyong*, or the "Doctrine of the Mean" (Gao et al., 2022). The term *zhongyong* is composed of two Chinese characters: *zhong* (中), translating to "middle" or "center" and denoting a midpoint that avoids extremes, and *yong* (庸), understood as "ordinary" or "common," which points to everyday practice or constancy. Together, they capture a philosophy of balance, moderation, and harmony. The essence of *zhongyong* is to maintain equilibrium in one's actions, thoughts, and behaviors, avoiding radical shifts or excessiveness (Gao et al., 2022). This wisdom is articulated in the classic Confucian text, *Zhongyong*, which asserts that the middle path should be pursued consistently (Gao et al., 2022). Even those of renown are in error if they deviate from this balanced course.

Applying this principle to our digital engagements, *zhongyong* advises us to use technological advancements judiciously. While the digital landscape can be captivating, the "Doctrine of the Mean" reminds us of the importance of harmony in our interactions. We should not become excessively reliant on or engrossed in digital tools but should incorporate them in ways that harmoniously enhance our overall well-being and lived experiences. As we traverse the vast expanse of the digital realm, the timeless guidance of *zhongyong* encourages a balanced and moderate approach, ensuring that our virtual interactions complement, rather than overshadow, our tangible realities.

The Metaverse as a Translanguaging Space

Wei (2011) first coined the term "translanguaging space" to describe both the act of translanguaging and the space that is created through translanguaging. In this space, individuals with diverse linguistic and cultural repertoires feel comfortable creatively shifting between these repertoires and asserting their complete linguistic identity. Building on Li's work, which primarily focused on translanguaging in educational contexts like universities or weekend Chinese schools, subsequent studies have explored various types of translanguaging spaces. For instance, Hua et al. (2017) discuss the multimodality of translingual space in a Polish shop in London, while Mazzaferro (2018) presents examples of different settings where translanguaging is applied in daily practice. More recently, Kiaer et al. (2022) examined a range of translanguaging spaces, including personal, philosophical, and playful spaces. The metaverse offers a space that might be seen as a refuge for individuals from various backgrounds. Within its boundaries, there's potential for these individuals to interact and share their unique insights, while building connections grounded in mutual respect. For those who might be socially vulnerable or marginalized in the physical world, the metaverse could provide a platform to foster understanding and solidarity. However, it's essential to approach it with caution. Despite its promise, the metaverse demands a sense of responsibility from its users to maintain respect and inclusivity. Every participant has a role in ensuring that the environment remains conducive to open dialogue and mutual respect. In this way, the metaverse can act as an alternative avenue for interaction, but its potential benefits must be weighed against the responsibilities it demands.

Translanguaging Competence in Metaverse

While moderating digital realms and striking a balance between physical and digital exposure is essential, what is paramount is how we synthesize and traverse these diverse realities. In our rapidly evolving landscape, the emerging generation isn't merely tasked with balancing screen time with tangible experiences, but also with decoding the intricate semiotic nuances that pervade each space. This skill set is imperative for all, especially the emerging youth, as it aids in effective expression and fosters solidarity among individuals. In particular, digital engagement empowers individuals to communicate beyond geographical boundaries, promoting a deeper connection and understanding regardless of physical distance. I've termed this pivotal skill as "translanguaging competence," a concept I introduced in 2023 (Kiaer, 2023b). In my book focused on children's language acquisition, I used "translanguaging competence" to elucidate young multilingual children's adeptness in modulating languages, necessitating distinct linguistic and cultural shifts. These children innovatively harness their multimodal semiotic tools, ensuring that their communication is not just socially attuned but also culturally and emotionally enriched. At its core, this competence is about fine-tuning one's language based on the listener's needs, effectively recalibrating semiotic elements to foster both efficient and empathetic communication. Drawing inspiration from Steven Pinker's (1994) *Language Instinct*, Li (2016) introduced the idea of a "translanguaging instinct" to account for the versatile linguistic applications by multilinguals. While there's a similarity between "translanguaging instinct" and "translanguaging competence," I've chosen to align with the latter. It resonates more closely with the innate properties of human language, reflecting conventions rooted in Chomskian linguistics.

Translanguaging competence is an expansion on the principle of translanguaging, which describes how multilingual individuals navigate their language and literacy. They don't merely switch between languages; instead, they seamlessly and creatively blend elements from two or more languages in their communication (Garcia & Li, 2014). Translanguaging is a dynamic process of meaning-making, whereby individuals use various semiotic resources across language and cultural boundaries (Lewis et al., 2012). In our multicultural societies, this practice is increasingly becoming the norm. An analogy I often employ is cooking. Consider the dish you will prepare tonight; it may incorporate ingredients that reflect your cultural heritage, but often there is a fusion. We mix and innovate, transcending boundaries, catering to our taste buds constantly. Spices traditionally used in Indian cuisine, for example, may have once had a clear connection to Southeast Asian food, but now that connection is more loosely defined, allowing us to use them as we please, depending on the occasion and the individuals we are catering to. Flexibility and hybridity lie at the core of this linguistic practice, although it may sound novel since the term has gained popularity only recently. However, throughout our history, we have engaged in this practice.

Further, I extend the concept of translanguaging beyond physical realities and into the metaverse—the hybrid space where digital, virtual, and physical realities coexist. In the realm of the metaverse, the sheer volume and diversity of each individual's semiotic repertoire seem almost infinite. To navigate this vast expanse, a particular digital competence becomes paramount: the ability to identify and adapt to the most suitable medium and reality for any given purpose. Translanguaging competence as such may very well be the most essential competence for our future endeavors. Expanding on this idea and building upon the concept of translanguaging, I propose in this book the new term "transversing"—the act of moving through different realities, both virtual and physical, or even creating new spaces and media to seamlessly express and share experiences with others.

Linguistic Diversity at Its Peak

The metaverse, in its expansive digital environment, provides a platform that reflects the multifaceted nature of human expression. Take, for example, two individuals interacting within a shared virtual space: one might choose to represent themselves with an avatar reminiscent of a favorite fantasy novel, while another might opt for an appearance grounded more in realism. Their choices of virtual locales might differ as well: one might favor bustling virtual cities and the other serene digital landscapes. Moreover, the platforms and tools they use to engage with these virtual realities could vary widely. One user might lean toward platforms optimized for intricate design, while another might prioritize ease of use or social interaction. Just as two chefs can take the same ingredients and produce vastly different dishes, individuals in the metaverse reshape and interpret their realities in distinctly personal ways. As a result, we observe an environment where linguistic diversity is at its zenith. The act of translanguaging, or fluidly navigating between languages, becomes even more complex and nuanced. In such a setting, it's not merely about mixing languages but also blending individual interpretations, preferences, and experiences. This leads to a situation where it's less about a collective language and more about individual "idiolects." The metaverse, therefore, becomes a haven for such individualistic expression, almost making it a digital embodiment of linguistic plurality.

Linguistic Anarchy in the Metaverse: Too Much Freedom?

Indeed, the metaverse, a complex digital universe, offers boundless avenues for linguistic expression. Within this domain, conventional language constructs melt away, making room for an emerging age of idiolects and sociolects that redefine our modes of communication. Emojis, memes, symbols, and a fusion of linguistic elements from a spectrum of languages merge to form a dynamic tapestry of expression. Notably, everyday individuals frequently employ a mix of languages, often in romanized forms, further enhancing this tapestry. Moreover, the rise of "image words" like emojis and memes, which are inherently transnational, adds another dimension to this linguistic evolution. These visual forms of communication transcend borders and cultures, amplifying the universal appeal of such expressions. As we delve deeper into this unrestricted linguistic realm, a pressing question arises: Are we teetering on the edge of linguistic anarchy? This shift is not merely confined to the metaverse. Signs of this linguistic transformation are evident in our daily digital interactions. Social media platforms and instant messaging apps have already witnessed a departure from the strictures of traditional grammar. Furthermore, these platforms may exert more influence on individual language use than even established national language authorities.

English in the Metaverse: Still the Lingua Franca?

Currently, English stands as the most widely spoken language worldwide, serving approximately 2.3 billion individuals as either their primary or secondary tongue. Historically, it has acted as a bridge between diverse cultures, particularly in influential sectors such as international business, diplomacy, entertainment, and education. Its vast influence has marked it as the world's "lingua franca," an unrivaled force in global communication. However, as we delve deeper into the digital and metaverse eras, the very fabric of linguistic communication is transforming. While the importance of English remains evident, especially with most online content primarily presented in this language (Petrosyan, 2023), the horizon showcases a future that is less predictable. Though it seems plausible that English will retain its dominant role as a common language platform, the emerging landscape hints at a richer tapestry. The 2023 report from the British Council by Patel et al. underscores this point. While it acknowledges English's central role in the global stage—from high-level business communications to sectors like retail and tourism—it also suggests a burgeoning evolution in our communication landscape.

With the rise of the metaverse and the increasing emphasis on digital communication, it's becoming clear that our future won't just be shaped by traditional languages and their alphabets. Instead, a more inclusive and diversified linguistic environment is unfolding. Alongside English, we're likely to witness a resurgence of other languages, and not just in the spoken or written form. Image-based visuals, symbols, and diverse signs are poised to become integral to our communication strategies, cohabiting seamlessly with more traditional linguistic forms. Moreover, the synthesis of technology and language learning points toward a more integrated future. While digital tools can occasionally create barriers, they also usher in a realm of possibilities, introducing a blend of human touch with digital enhancements. This is evident in the report's emphasis on the invaluable role of human teachers, even in an age dominated by AI and auto-translation. Furthermore, the future of English-medium education appears fluid. While it's growing as a preferred medium in higher education, there's a tangible reluctance in non-native regions. This hesitancy, combined with the evolving nature of global communication, signals a future where English is just one of many vital players.

The prominence of English isn't just a testament to native speakers. Its influence extends deeply into non-English-speaking nations. These countries, recognizing the weight English holds in global discourse, have incorporated it extensively into their education systems, businesses, and even pop culture. Japan's love for English loanwords, France's adoption of English in their start-up ecosystem, and the popularity of English Medium Instruction (EMI) in various parts of Asia are testimonies to its global acceptance. In the metaverse, users from these countries often default to English as the primary mode of communication, making the language a universal connector once again.

While the significance of English is hard to ignore, it would be an oversight not to recognize the burgeoning of its regional variants. Singlish in Singapore, Hinglish in India, and Spanglish in the United States are not just colloquial anomalies but are finding their rightful space in the broader English lexicon. As the metaverse promotes deeper cultural exchanges, these regional variants gain more visibility and acceptance. Instead of diluting the influence of standard English, they enrich it, adding layers of cultural depth and making the language more inclusive. This diversification could, in fact, be the key to its longevity.

A Hugely Diversified Future?

With the onset of the metaverse age, we're not merely looking at a digital landscape dominated by one language. Instead, we're witnessing the blossoming of a more diverse linguistic landscape, with English at its core, but surrounded by its many offshoots and regional flavors. This doesn't detract from its power but rather accentuates it. While the metaverse may introduce various languages and cultures more profoundly, English, with its vast spectrum of regional variants, is poised to remain a significant lingua franca. Its evolution from a dominant global language to a diversified yet unifying force epitomizes the inclusive potential of the digital age.

More Inclusive Communication

Communication in the metaverse has the potential to revolutionize interactions by fostering diversity and inclusivity while mitigating biases. Unlike traditional face-to-face conversations, immersive communication in the metaverse enables individuals to engage without the constraints and preconceptions associated with physical presence. By embracing virtual avatars and transcending our physical identities, we can cultivate an environment that celebrates and respects the richness of diversity. The metaverse offers a unique opportunity to bridge divides, dismantle barriers, and cultivate a sense of belonging for all participants, regardless of their background or physical attributes. Through immersive communication, we can redefine the way we connect, communicate, and coexist, establishing a digital world that is more equitable and inclusive. In theory, within the metaverse, one can discover a space where concerns about gender, race, and other factors diminish, promoting a sense of equity and acceptance. However, it is crucial to remain vigilant, as the metaverse also holds the potential for social prejudices to emerge and instances of abuse to occur. The metaverse represents a realm of immense potential, although it is not without its imperfections. Its gamified appearance makes it appealing and accessible to newcomers, yet it may undermine the gravity of certain discussions. Additionally, compared to traditional Zoom meetings, it may lack the authenticity and genuine emotional connection. The high cost of tools such as headsets poses a challenge for the widespread adoption of a fully immersive metaverse. However, it is important to note that as technology continuously evolves, these limitations may change over time.

4

Immersive Communication

The era of virtual migration has arrived, fundamentally reshaping the way we communicate. Whether it's through the convenience of our smartphones, the seamless integration of smartwatches, or the assistance of familiar AI companions like Siri or Alexa, our conversations and interactions are increasingly intertwined with the digital aids of our choice. The online world we engage with, presented to us through screens, encapsulates the essence of a virtual realm, liberated from physical constraints. This shift to the virtual realm, while already in motion, was further propelled by external factors. The Covid-19 pandemic served as a catalyst, driving us deeper into these digital spaces. It showcased the potential of the virtual domain, demonstrating its capacity to support work, recreation, and social interactions. As of the Web 3.0 era, the digital realm ceased to be just a place to visit; it evolved into a platform for creation. Here, every individual can be a participant and a creator, molding content and even reshaping platforms. This cocreation leads to organic innovation, making interactions within this realm even more immersive and captivating. A shining beacon of this shift toward a cocreative, super-digital society can be seen in how we adapted to tools in times of need. Take Zoom, for instance. When it skyrocketed to popularity in March 2020, even its developers couldn't fully envision the myriad of ways it would be employed. Yet users, in their adaptive ingenuity, found innovative purposes for the platform. Zoom wasn't just for meetings. It became a venue for parties, virtual games, children's sleepovers, and movie nights. This instance underscores the core idea: when given a digital space, users don't just occupy—they innovate. In this present age, it's not just about typical face-to-face spoken communication, but also about how we interact with others and various services in the digital realm. In this space, communication encompasses more than speech; it includes a vast array of our daily activities, such as gaming, using streaming platforms, and even shopping. Our journey toward virtual migration isn't a passive move to a new domain. It's an active evolution, fueled by collective creativity and the boundless potential of digital platforms.

Birth of Emojis: Transformation of Emoji Culture

The birth of emojis was a response to the constraints of text-based communication. Their trajectory has been influenced by the emergence of memes and the quest for more diverse, individualized modes of expression. These shifts underscore the ever-evolving nature of human communication as it adapts to the nuances of digital environments. Initially, emojis emerged as a solution to the inherent limitations of texting, infusing conversations with a newfound expressiveness that once eluded typed words. The realm of texting, although convenient, grapples with conveying emotions and attitudes effectively due to its inherent constraints. To bridge this communication gap, a diverse set of expressive tools was introduced, encompassing emojis, emoticons, memes, and avatars. However, they failed to fully fulfill the yearning for a more comprehensive communication that could encapsulate the depth of real-life interactions. The early days of emojis were characterized by preset options, which often struggled to encapsulate the full spectrum of diversity in human expression. The emoji culture has rapidly expanded, outgrowing the confines of standardized sets. The previous norm of relying on a limited collection of emojis has given way to a more personalized approach. The once predominant yellow emojis are no longer the only option. People craft their own emojis and memes. This trend is poised to continue, promising even further diversification and personalization of memes.

The realm of emojis is undergoing a significant transformation with the introduction of immersive functions. Emojis are transitioning toward a more individualistic paradigm. People's innate need for self-expression defies the confinement of a limited array of yellow emojis. The trend of crafting custom emojis and memes is experiencing exponential growth, promising an even richer and more varied landscape. In parallel, the concept of memes, initially introduced by Richard Dawkins (Dawkins, 1976), has risen as a cornerstone of culture. Much like genes in biological evolution, memes exert considerable influence on cultural dynamics, particularly in our rapidly evolving world. Memes operate as infectious ideas, propagating through imitation from one individual to another. This phenomenon mirrors the propagation of living organisms, utilizing human brains as vehicles of dissemination, akin to viruses. Nonetheless, the dynamics of memes are multifaceted. In contrast to genes, which generally bestow advantages upon their hosts, memes don't consistently yield benefits to those who adopt them. Dawkins underscores that the evolution of memes doesn't always align with believers' interests, in contrast to the role of gene evolution in biology. The concept of memes, introduced by Dawkins, draws intriguing parallels between cultural and biological evolution. However, this idea remains in its nascent stages, akin to a seed necessitating further nurturing through time and research, akin to the maturation of genetic theories.

Meme Culture: Does Digital Reflect Physical?

The culture of reinterpreting and recreating existing materials, combined with prevalent sharing, is distinctly evident in today's meme culture. Memes act as a form of AR for the masses, allowing individuals to project their interpretations, humor, and societal critiques onto existing digital canvases. As such, they're a significant part of the digital experience for many. Yet, the very act of creating and sharing memes brings to light profound challenges related to copyright. In the world of memes, where content is endlessly repurposed, remixed, and reshared, the boundaries of original ownership and rights often become blurred. In the physical world, the act of replicating another's creation without permission is clearly a breach of copyright, but in the digital realm, particularly in the world of memes, such boundaries are not always acknowledged or respected. Limor Shifman's (2013) deep dive into this digital phenomenon is enlightening. Her meticulous analysis underscores the multifaceted nature of memes and their unique position at the crossroads of creativity and potential copyright infringement. With each meme's transformation, a potential challenge to traditional copyright norms also arises. The phenomenon of memes—reinterpreting, reimagining, and then sharing cultural touchstones—shines a spotlight on the challenges of intellectual property in the digital age. Memes, which dominate platforms like Instagram, Twitter, and Facebook, are emblematic of a larger trend where the boundaries of ownership and copyright are tested daily. For many internet users, the act of reshaping and redistributing content feels organic and harmless, often serving as a mode of social commentary, humor, or even critique. The quandary, however, extends beyond just memes. As augmented and virtual realities become more integrated into our digital lives, the conundrum grows. How can we transpose the clear-cut laws of the physical world, where ownership and rights are often demarcated by tangible boundaries, onto the fluid landscapes of the digital and virtual domains? Is it possible for regulations crafted for tangible creations and interactions to be directly and fairly applied to these ever-shifting digital constructs?

Memes, being at the heart of this debate, exemplify the profound challenges at hand. Their widespread creation, alteration, and dissemination on social media platforms might be seen as a community's relaxed approach or even disregard for traditional copyright norms. It's crucial to understand that this isn't necessarily borne out of a mass intention to violate intellectual property rights. Instead, it could be a result of the digital age's inherent culture, where sharing, remixing, and reimagining content are seen as standard, even commendable, behaviors. Navigating this new frontier necessitates a comprehensive reevaluation of how we understand and enforce copyright in the digital realm. What's abundantly clear is that norms and regulations from the tangible world don't always neatly fit into the virtual one. As we grapple with these challenges and ethical considerations, it's evident that our legal and moral frameworks need to evolve. They must adapt, if not undergo a significant transformation, to adequately address the unique challenges and nuances presented by our evolving digital landscapes.

Navigating Intellectual Property, Copyright, and Emerging Challenges in VR Spaces

The domain of intellectual property, including copyright, encounters newfound complexities within the realms of AI and VR. As technology advances, it often outpaces the legal framework designed to govern it, leaving us with intricate questions. Take, for example, the creation of Memoji using smartphones—an everyday practice for many. Surprisingly, it's not entirely clear who truly owns these digital likenesses. Venturing into the expansive world of virtual reality amplifies these complexities. In the physical world, actions and ownership are traceable, but within the virtual expanse, millions interact without clear attribution. This ambiguity extends to issues of surveillance and security, introducing entirely unprecedented challenges. The surveillance landscape in VR is vast and largely uncharted, raising questions about privacy, data protection, and the potential for misuse. The evolving interplay between AI, VR, and IP necessitates a dynamic and adaptable approach to address these emerging issues, ensuring a balance between innovation and protection in this always-changing digital frontier.

Metaverse: Complementing the Online and Offline World

The evolution of online communication has consistently demonstrated the inherent challenges of bridging the gap between offline and online interactions.

Traditional online meetings, while invaluable in certain contexts, can't replicate the full spectrum of human communication. The nuances of spontaneous conversation, the subtleties of nonverbal gestures, and the intimacy of haptic touch remain elusive. Emojis, despite their widespread use, offer only a limited palette of expressiveness. It's neither natural nor easy to ask questions or provide feedback in most online platforms. In my previous work, *Pragmatic Particles* (Kiaer, 2020), I proposed the 3E model (efficiency, expressivity, and empathy) as the driving forces behind effective human communication. While efficiency pertains to the sheer volume of information conveyed, the true essence of communication extends beyond mere information exchange. The depth of expressivity and the richness of relationship-building are where conventional online methods often fall short compared to face-to-face encounters. Online meetings greatly aided global communication during the pandemic, serving as invaluable tools to bridge physical distance and sustain both personal and professional connections. They've been revolutionary, to say the least. However, there's an emerging sentiment: a craving for something more immersive and interactive than what's provided by standard video call platforms. Not just gamers but professionals and everyday individuals are seeking richer, more personalized experiences.

Enter the metaverse, augmented by the power of VR. Within the metaverse, employers are presented with an unprecedented opportunity to enhance telework practices. More than just a 3D video call, it offers an environment where understanding a colleague's facial expressions and body language becomes instinctual, making communication more intimate and effective. In this virtual space, even presentations can be reimagined, delivered with a creative flair that transcends traditional slide decks. However, the promise of the metaverse doesn't come without its challenges. As businesses extend into these virtual territories, there's a pressing need to reconsider tax, social security, and even immigration rules for workers operating in countries they don't physically inhabit. Employers must also be vigilant, drafting new codes of conduct that address metaverse-specific behaviors, from avatar appearances to virtual communication etiquette.

Several scholars have already touched on the harmonious relationship between online and offline interactions. Pettersen underlines the significance of physical spaces in nurturing interactions and fostering a sense of meaningfulness and insights that transcend mere virtual collaborations. Zhang and Venkatesh (2013) identify a synergy between virtual and physical communication networks, observing their combined effect on job performance. Kent et al. (2019) explore the relationship between in-person social capital and virtual interactions, noting a beneficial linkage between the two. In short, the metaverse isn't an entity that stands isolated from our current modes of communication; it's the next stage in the evolution of communication. As Croes and Antheunis (2021) and Angelopoulos and Merali (2015) have noted, while online communication platforms like the metaverse offer unparalleled advantages, they cannot and should not aim to replace offline communication. Instead, they should strive to enhance and complement it, crafting a future where virtual and physical realities harmoniously coexist.

Gesture Matters

Face-to-face communication is a richly nuanced, fully embodied interaction that doesn't just rely on the literal meaning of spoken words. Nonverbal cues, such as facial expressions and tone of voice, are crucial in conveying emotions and attitudes. Albert Mehrabian's (1972) pioneering research in this field brought to the forefront the profound influence of these cues, especially when there might be a dissonance between spoken words and nonverbal signals. Based on experiments from the 1960s, Mehrabian (1972) formulated the 7-38-55 percent communication rule. In face-to-face exchanges, words are responsible for only 7 percent of the overall message's impact, whereas tone and intonation contribute 38 percent. Surprisingly, a significant 55 percent is derived from nonverbal behaviors, with facial expressions at the forefront. Though this rule might not be universally apt, it emphasizes the vital importance of nonverbal communication. Mehrabian's research experiments have consistently demonstrated a key insight: In emotional communication, facial expressions and tone of voice often have a more profound impact than the literal meanings of words. However, Mehrabian's studies have faced critiques. Some point toward the limited sample sizes he used, potential gender biases, and the challenges of applying his findings to broader, real-world communication contexts. Still, the breadth of Mehrabian's impact is undeniable. His insights into nonverbal communication dynamics have been applied across diverse sectors.

Impact on Understanding Offline Interactions

The effects of online communication on offline interactions and interactional skills in general have also been widely studied. Ruben et al. (2021) investigate the effect of communicative technologies on interpersonal communication. Their results showed that passive usage of such technologies is likely to help in developing nonverbal decoding skills to enhance in-person communication, while active usage of said technologies results in decreased nonverbal decoding accuracy, which could have negative effects for in-person communication. Tricia Jones, a professor at Temple University, posited that the pandemic has led to a greater reliance on technological devices for communication, resulting in decreased exposure to in-person nonverbal communicative cues (Oputu, 2020). This, Jones says, has led to a new norm or social etiquette, whereby people are less adept at interpreting and using such cues in communication. Instead, they resort to explicit verbal ways of communicating one's intentions (Oputu, 2020). Bauerlein (2009) also echoes a similar view, taking a more pessimistic stance by postulating that generations of digital natives will have decreased nonverbal literacy, possessing "adroitness at the keyboard," but a diminished ability to "'read' the behaviour of others." York (2022), on the other hand, discusses how the shift toward remote working has necessitated the presentation and interpretation of nontraditional forms of nonverbal communication, such as making eye contact with a camera instead of a face during a video call when it is one's turn to present, or how turning the camera on or off could be an issue of politeness. Kang (2007) observed that online communication has two differing psychological outcomes: While it can function as a tool for offline interaction and lead to positive outcomes, like increased happiness and reduced depression, its disembodiment effect (i.e., a sense of detachment from the physical body or reality) can negatively influence psychosocial well-being and social support. Nevertheless, technologically mediated communication has its own limits. Brownell (2020) notes that it can be difficult to comprehend such forms of communication due to the lack of nonverbal cues that bolster one's understanding of the content being communicated. She elaborates on this, stating that "when you communicate through Skype or Zoom, you may feel you are 'connected,' but there is still a perceptual distance that impacts your behavior and response in ways that make understanding more difficult and more uncertain" (Brownell, 2020, p. 67). Typical video conferencing platforms often fall short in offering immersive and interactive features, making it challenging to capture the dynamic nature of human communication. Nonverbal cues and gestures play a significant role in interactions, but they're difficult to convey effectively through standard video conferencing. Additionally, the experience can be further diminished when someone simply turns off their video, eliminating any visual communication altogether.

The Power Dynamics of "Mute"

Video-conferencing platforms, such as Zoom and Teams, have rapidly become essential tools in the world of digital communication, especially in the realm of education and professional meetings. One of the most recognized features of these platforms is the "mute" button, a seemingly simple tool with complex implications. At face value, the "mute" button serves a pragmatic purpose: ensuring clarity by reducing background noise, preventing interruptions, and creating an orderly environment, especially in larger meetings or lectures. In scenarios where a single speaker must address an audience without distraction, this feature is undeniably beneficial. However, when we delve deeper into more interactive settings like debates, discussions, or collaborative meetings, the dynamics of the "mute" feature take on a more nuanced role. It can become a tool of power, subtly enforcing hierarchical structures. The person or persons with administrative rights, often the hosts of the meeting, wield the ability to silence participants. In the broader context of digital communication ethics, it's essential to critically evaluate even the most benign-seeming tools. The "mute" button, while practical, is a testament to the subtle ways technology can shape and influence power dynamics. While this may often be used with the best intentions—for instance, to manage crosstalk or background disturbances—it can also be utilized, intentionally or not, to control narratives and dictate the flow of conversation. This brings to light an essential concern: Does the "mute" button inadvertently perpetuate a system where the powerful control the powerless? In discussions, especially those that touch on controversial or contentious topics, the ability to mute can suppress dissenting voices, hinder open dialogue, or even be used to dominate the conversation. Moreover, the act of muting, particularly when done without the consent of the individual being silenced, can be disempowering. It sends a clear message about who is in control and can discourage active participation, creating an environment where participants may feel hesitant to share their views.

What Is Immersive Communication?

The *OED* defines "immersion" as "Absorption in some condition, action, interest, etc.". The desire for immersive communication is not a new concept. Throughout history, people have sought ways to connect with each other on a deeper level. As early as 1964, AT&T demonstrated their videophone called the "Picturephone" at the New York World's Fair (Apostolopoulos et al., 2012). Immersive communication encompasses a multichannel, multisensory experience, engaging both sight and sound. It involves nonlinear and exploratory interactions, granting the audience the freedom to choose how they want to interact with content and take control of their experience. Immersive communication refers to a form of communication that deeply engages individuals in a simulated or virtual environment, creating a sense of presence and heightened sensory experience. It utilizes VR, AR, MR, or other immersive platforms to enhance the interactive and immersive nature of the communication process.

This form of communication becomes even more crucial as we encounter a growing number of digital experiences, as existing platforms often lack the desired level of interactivity and a complete sense of embodiment. The limitations inherent in traditional communication methods have kindled a longing for immersive experiences, which is further propelled by continuous advancements in technology. These technological strides incessantly push the boundaries of what is conceivable, enabling the creation of increasingly immersive and captivating encounters. Consequently, there is a burgeoning demand for enhanced modes of communication that effectively bridge the digital realm and the physical world. The metaverse offers more than a mere rental of its platform or physical space; it comes bundled with a myriad of digital reinforcements. Indeed, the metaverse necessitates our coexistence with AI, leaving us no choice but to live alongside it. Immersion becomes our reality, and the path to immersion is often guided by AI. Our interactions and conversations within the metaverse are shaped by the extent to which AI and other technological entities permit and enable us. In this sense, the metaverse is far from a neutral space, challenging our initial hopes and aspirations.

Immersive Communication: Efficiency

Immersive communication significantly enhances human interaction. As previously mentioned, in my earlier work, *Pragmatic Particles* (Kiaer, 2020), I introduced the 3E model and also considered how true communication goes beyond mere data transfer. Relying solely on letter-bound verbal communication can lead to misunderstandings; it's challenging to convey emotions and nuances. Additionally, the barrier of language can further complicate understanding. However, by incorporating visuals and sounds into our conversations, not only is understanding enhanced, but many language barriers are also transcended. This approach is particularly relevant for our emerging generations. Gen Z and Gen Alpha, in particular, are more attuned to processing multimodal information as opposed to just text-based data. This signifies that immersive, multisensory communication isn't just a current trend, but the way forward.

In the digital age, the allure of multimodal media isn't confined just to the youth. Across all demographics, there appears to be a marked shift toward this versatile form of communication, underpinned by the promise of efficiency and the speed it offers in our hyper-accelerated world. Multimodal media captivates by activating multiple senses at once. It combines text with visuals, sound, and occasionally tactile stimuli, culminating in a comprehensive experience that aids comprehension and memory retention. This convergence of media types taps into the brain's capacity to process diverse inputs concurrently, fostering a more profound and enriched engagement with the content. Consider the contemporary manner in which we consume news. The seamless integration of infographics, videos, and auditory elements caters to diverse learning predilections, appealing to visual, auditory, and kinesthetic learners in tandem. This evolution is further exemplified by platforms like YouTube. Where once extended videos reigned supreme, there's a discernible pivot to "Shorts"—crisp, condensed visual narratives. This shift underscores a burgeoning preference: In our fast-paced world, there's a relentless quest for expeditious yet substantive content consumption.

Immersive Communication: Expressivity

Immersive communication in the metaverse can also provide a richer, all-encompassing platform that makes human interaction deeply engaging and expressive. This enhanced form of communication leverages the depth of semiotics, the study of signs and symbols, allowing users to fluidly integrate elements from both the physical and digital realms. By doing so, individuals can tap into a vast potential for expression that is uniquely tailored to their persona, surpassing traditional digital interactions. One of the most transformative aspects of this is the ability to transcend the barriers of language and culture. With the metaverse's tools, linguistic and cultural differences can be effortlessly bridged, enabling universal understanding. All these promising avenues of communication are expanding in tandem with the rapid advancements in digital technology, suggesting that as our tools evolve, so will our ways of connecting with one another. The true marvel of this space lies in its inclusivity and expansiveness. For those who might find it challenging to articulate themselves in the physical world—whether due to societal norms, personal anxieties, or other barriers—the metaverse offers a sanctuary. Here, individuals can craft, modify, and redefine their modes of expression, pushing the limits of what was once thought possible. However, this vast sea of possibilities also has its depths. While it offers liberation to many, it might prove daunting for others, potentially leading to feelings of disconnection. Nevertheless, the overarching potential of the metaverse remains captivating. As our digital tools evolve and intermingle with our physical realities, they are forging an unprecedented frontier for communication. In this evolving domain, our imagination stands as the only conceivable boundary to expression.

Immersive Communication: Empathy

Empathy, often considered the cornerstone of genuine human connection, finds its place even in the vast expanse of the metaverse. The metaverse, with its immersive communication tools, offers an extension of our human need for empathy and connection. Though it's a digital frontier, it echoes the age-old quest for understanding and solidarity. The challenge, and opportunity, lies in navigating its intricacies to foster genuine and lasting connections. Just as in the physical realm, the digital spaces of the metaverse are ripe for understanding, feeling, and connecting with others on a deep and emotional level. Building solidarity stands out as one of the primary purposes of immersive communication. The digital age, punctuated by the metaverse's ascendancy, has presented a unique landscape for individuals to share, relate, and form bonds. While for some, especially the younger generation, it's a continuation of the norms set by social media and life logging, for others it's a novel avenue for making connections.

As in our tangible world, the metaverse offers the promise of forming friendships, building communities, and fostering a sense of belonging. People from diverse backgrounds and ages are discovering that the digital space can, quite surprisingly to some, resonate with the warmth and understanding found in face-to-face interactions. The essential goals of human communication—to inform, to persuade, and to entertain—seamlessly transition into this new realm. However, while the potential for empathy and solidarity in the metaverse is evident, it's also crucial to acknowledge its dual nature. For every individual who finds comfort and connection, there might be another who feels adrift in the digital vastness. The very platforms that weave communities for some might inadvertently create barriers for others.

Born to Immerse and Interact

While visiting Somerset House with my eleven-year-old daughter, we paused before Édouard Manet's painting, *A Bar at the Folies-Bergère* from 1882. She placed her toy next to the painting and captured the moment with a photograph. This act brings forth a pressing question: While the photograph clearly belongs to my daughter, does it also hold a claim from Manet due to his work's inclusion? If so, is this an infringement on copyright? This becomes even more complicated when considering the sheer number of people sharing similar content daily on social media. If such acts are indeed breaches of copyright, then are millions engaging in illegality each day?

Before the surge of digital technology and social media, most people were largely recipients of mainstream culture, with limited avenues to actively participate or reshape narratives. Now, there's a heightened, almost insatiable desire not just to be a spectator but to be an active participant. People want to reinterpret, to place themselves within the stories, and to be a part of the broader narrative. This immersive and interactive desire has been supercharged by digital advancements. The emergence of social media platforms initially bridged this gap, giving everyone a platform to express, share, and reshape their narratives. Now, the metaverse—a hybrid space blending the digital and physical realms—is the frontier where this desire finds its most expressive outlet. It's where we constantly encounter individuals striving not just to exist but to actively participate and shape this evolving world. With the dawn of the metaverse, even traditional experiences like viewing art in a gallery are undergoing a paradigm shift. It's no longer about mere observation. Instead, it's about an active engagement, a deep-rooted desire to immerse oneself, and to be a living part of the experience. The inclination to immerse ourselves in the metaverse, seamlessly transitioning between the digital and physical worlds, has become deeply embedded in our daily lives.

Immersive Consumers

Businesses are increasingly recognizing the value of these immersive platforms, while consumers are demonstrating a growing desire for richer, more enveloping digital experiences. The horizon of immersive desire is, undoubtedly, expanding. The burgeoning interest in immersive experiences is underscored by the recent projections in the virtual reality market, which estimate that the combined VR and AR market could reach $1.5 trillion by 2030 (PwC, 2019). A study titled "Virtual Reality Market with COVID-19 Impact Analysis by Offering (Hardware and Software), Technology, Device Type (Head-Mounted Display, Gesture-Tracking Device), Application (Consumer, Commercial, Enterprise, Healthcare) and Geography—Global Forecast to 2025" delves into this growing realm, providing an analysis that encompasses various facets, including hardware and software offerings, technologies, device types like head-mounted displays and gesture-tracking devices, as well as diverse applications spanning from consumer to healthcare sectors (MarketsandMarkets, 2020). The findings of this report are particularly illuminating. It estimated that by 2025, the virtual reality market is anticipated to swell to an impressive USD 20.9 billion, charting a compound annual growth rate (CAGR) of 27.9 percent over the forecast period. This rapid expansion not only underlines the immense potential and commercial viability of virtual reality but also signals a clear shift in both corporate investment priorities and consumer appetites.

Google Earth

On the cusp of VR is an immersive online tool used by many across the globe: Google Earth. Google Earth is a geospatial tool developed by Google, providing users with satellite imagery, aerial photography, and various topographical details from locations around the world (Weng, 2012). It facilitates a digital representation of the Earth's surface, allowing users to explore places ranging from vast geographical regions to specific urban environments. In the educational context, particularly within classrooms, Google Earth serves as an instrumental resource. For instance, in a geography lesson focusing on topographical features, educators can employ Google Earth to provide students with a visual representation of mountain ranges, river systems, and deserts (Demirci et al., 2013). This digital visualization offers an enhanced understanding of spatial relationships, geographical scale, and the diverse terrains that constitute our planet.

Beyond geography, the utility of Google Earth extends to subjects like history and literature. History educators can use the tool to contextualize significant events within their geographical settings (Tarr, 2006). By doing so, students gain a spatial understanding of historical narratives, enhancing their grasp of events in relation to their locations. Similarly, literature instructors can transport students to the settings of novels or poems, adding a layer of spatial context to literary exploration (Boss, 2008). One of the key features of Google Earth is its capability for customization. Educators can annotate specific locations, designating points of interest pertinent to their lesson plans. Moreover, students can engage in independent projects, using the platform to create presentations that are geographically informed. Google Earth, with its vast database of geospatial information, stands as a valuable asset in the academic realm. Its ability to visually represent and contextualize geographical information provides educators and students with a multifaceted tool to enhance learning experiences. As technology continues to develop, we will see its incorporation into more and more spheres of our lives, from education to entertainment and beyond.

Further, the integration of immersive AI technologies with platforms like Google Earth can offer an unprecedented level of detail and interactivity, transforming how we understand and engage with our planet. Imagine an enhanced Google Earth experience where artificial intelligence algorithms process and augment real-time data streams from satellites, sensors, and user-generated content. These could provide contextual information, simulate future environmental changes, or even offer predictive analytics for natural disasters; AI could work in tandem with computer vision to identify and label geographical features, landmarks, or points of interest dynamically. This would not only be a valuable tool for educational and research purposes but also could have practical applications, such as in urban planning, conservation efforts, and emergency response planning. Such an advanced, AI-driven version of Google Earth would essentially serve as an ever-evolving, interactive digital twin of the real earth, offering both visual and data-driven insights that are far beyond what static maps or traditional earth observation methods can provide. This kind of synergy between AI and immersive technologies has the potential to profoundly impact multiple sectors, from education and research to governance and public safety.

Immersive Entertainment

In the entertainment realm, during this modern age of digital fandoms, it's not just about loving a particular artist or group; it's about immersing oneself in a world, a narrative, and an experience that extends beyond the music. This phenomenon of creating intricate "universes" within fandoms is prominently observed in the K-pop industry—for instance, where imaginative storytelling and world-building have become essential tools for engaging fans and differentiating artists (Pratapa, 2021). Traditionally, the allure of pop music has rested largely on melodies, performances, and the charisma of artists. However, K-pop, known for its trailblazing approaches, has pushed the envelope by weaving intricate narratives into its musical offerings (Yoon, 2019). This method, while not entirely new, has been elevated to new heights, transforming the way fans engage with their favorite artists. The idea of creating a unique "universe" or backstory isn't just a creative endeavor; it's a strategic one (Jie, 2022). When the K-pop group EXO, managed by SM Entertainment, debuted with a storyline suggesting that its members were aliens from an "Exoplanet," fans were intrigued (Jie, 2022). This was not just another boy band; they had a narrative, a story that fans could follow, discuss, and invest in. It was an inventive way to solidify their brand identity in the saturated K-pop market.

This form of engagement is much more profound. When fans invest in a universe, they're not only waiting for the next song or album; they're anticipating the next chapter in a continuous saga. It gives fans additional content to dissect, fan theories to discuss, and more avenues to feel connected with the artists. Other K-pop groups, like aespa and BTS, have further expanded on this concept. While aespa navigates the blurred lines between virtual and reality, BTS's "Bangtan Universe" offers an intricate storyline that touches upon themes of youth, adversity, and growth (Jeong, 2020; MTV News Staff, 2021). This has opened new doors for fan interaction, where every music video, tweet, or concert can be a clue or a continuation of their respective universes.

However, this trend has been met with mixed reactions. Some industry experts see it as the future of fan engagement, believing that in an era defined by augmented reality, virtual worlds, and the metaverse (Young & Stevens, 2023), these universes cater to evolving fan expectations. On the flip side, there are concerns. Critics argue that an excessive focus on universe-building could detract from the primary musical essence of these groups (Jie, 2022). There's a risk of alienating the more casual listener or placing the artist in a narrative box, potentially limiting their musical evolution. In the grand tapestry of music history, K-pop's universe-building might be a relatively new thread, but its impact is undeniable. The phenomenon underscores a crucial aspect of modern fandoms: the desire for deeper, more immersive connections with artists (Giorgio et al., 2023). As the line between reality and the virtual realm continues to blur, these universes, borne out of creative storytelling, are set to play an even more significant role in defining artist–fan relationships in the digital age.

The growing influence of digital platforms on real-world, physical, and political events is hard to ignore. A notable example was when K-pop fans and TikTok users organized a campaign to claim tickets for Donald Trump's Tulsa rally with the intent of not attending, leaving numerous empty seats at the event. This coordinated effort underscored the power online communities possess to generate tangible, real-world outcomes, cutting across geographical and cultural barriers. Within these virtual spaces, the dynamic of parasocial relationships also takes on new dimensions. Fans transition from being mere spectators to emotionally invested parties actively engaged in the lives of their favorite artists. Platforms like Weverse are increasingly becoming catalysts for deepening the emotional bonds between fans and artists. Developed by Hybe Corporation, Weverse provides a variety of both free and paid content and acts as a direct channel for more intimate artist-to-fan communication. This increased emotional investment forms a strong foundation for collective action around diverse causes, ranging from social issues to political activism. Specifically, Weverse's multilingual support and global reach enhance its ability to inspire coordinated real-world actions on an international scale. As the digital landscape continues to evolve, it's likely that more platforms akin to Weverse will emerge, each with the potential to act as powerful agents for real-world change. Digital platforms are more than mere outlets for passive consumption; they are evolving into instrumental channels that can drive tangible changes in our physical and political landscapes.

Immersive Experiences: The Pleasure and Safety of VR

Virtual realities provide an avenue for unparalleled pleasure-seeking, a space where our wildest dreams become tangible, yet the risks are minimized. We're granted the freedom to explore, to push boundaries, and to experience the new, all while remaining in a controlled, safe environment. It's an evolution of entertainment, a blend of desire and innovation, rooted in our innate human longing for experiences that both thrill and protect. In the age of technology, the way we seek pleasure and entertainment has witnessed a profound shift. The emergence of immersive experiences, like those offered by VR, presents a tantalizing proposition: the ability to live out vibrant, compelling scenarios from the comfort and safety of our homes. These simulated worlds beckon with the allure of adventure and novelty while simultaneously ensuring that users remain insulated from the financial and physical risks often associated with real-world experiences. Consider the adrenaline rush of skydiving. Through VR, one can plummet thousands of feet, feel the wind rush by, and witness breathtaking vistas, all without ever boarding a plane or fearing the parachute won't deploy. This experience isn't merely a visual or auditory simulation.

In the vast expanses of the metaverse, a significant draw for users is the deep, immersive interaction it offers. The metaverse transcends traditional digital experiences, allowing users to engage in lifelike scenarios, conversations, and activities. This sense of being "inside" the experience, rather than just observing it, holds immense attraction. Such immersion goes beyond visual depth, resonating with our innate human desires to connect, explore, and feel. As the brain perceives these virtual interactions, mirror neurons come into play, making users feel as though they're genuinely living out these experiences (Portnoy, 2017). This intricate dance between technology and neuroscience provides not only a sense of escapism but also a genuine feeling of pleasure derived from the seemingly tangible connections formed within the metaverse's confines. Thanks to our brain's intricate wiring, particularly the presence of mirror neurons, users don't just see or hear these experiences—they feel them (Huntington, n.d.). Mirror neurons activate both when we perform an action and when we witness that action being performed (Acharya & Shukla, 2012). Mirror neurons are specialized cells in the brain that activate both when an individual performs an action and when they observe that same action being performed by someone else (Bates, 2009; Acharya & Shukla, 2012). Essentially, they allow us to "mirror" or internally replicate the observed activity, blurring the lines between self-experience and external observation. In the realm of VR, this means that watching your avatar navigate a VR space can evoke sensations as if you yourself are undertaking the action. This blurring of lines between the self and the avatar, between reality and the virtual, amplifies the intensity of the VR experience.

VR Vocabulary: A Lens into Virtual Interaction and Experience

The realm of VR, an intertwining of the real and the virtual, introduces a lexicon that often shifts away from our conventional definitions. This vocabulary isn't confined to just technical jargon but also expands into verbs that describe our experiences and interactions within these digital frontiers. When discussing the verbs that typically surface in VR contexts, we often encounter passive words such as "watch," "view," and "explore." While these verbs might suffice for a novice user, they don't encapsulate the dynamism and potential of VR. For instance, consider the term "collision detection." While it might hint at a mere physical crash in our day-to-day lives, in VR, it takes on a deeper role. It's about an avatar's interaction with a virtual object, ensuring the realism of VR experiences is maintained (Meseure & Kheddar, 2011). "Eye-tracking" in the daily vernacular might be about observing one's gaze. But in VR, it becomes a tool that optimizes visuals based on a user's focus (Matthews et al., 2020). The "Field of view (FOV)," usually linked with optics or camera perspectives in VR, determines the observable scope of the virtual environment, intensifying the immersion factor (Miller, 2021). While "gestures" in the physical realm might be as straightforward as a wave or nod, in VR, they evolve into commands. These can range from simple actions like pinching or pulling to more intricate choreographed sequences that dictate the VR narrative. The term "haptics" is elevated in VR. More than just tactile feedback, it's about sensations, from the feeling of raindrops to movement experiences, adding layers to the immersion "Locomotion" in VR isn't just about movement. Given the constraints of the physical environment, it often leans on simulated methods like teleportation to navigate expansive virtual landscapes (Matviienko et al., 2022). The word "presence" in VR holds special significance. Beyond its usual understanding, in VR, it paints the intense feeling of genuinely existing within a digital domain, where lines blur between the real and the simulated (Berkman & Akan, 2019). However, there's a growing desire to embrace a more active VR lexicon, especially in fields like education. Instead of merely "watching" a VR lesson, why not "interact," "collaborate," or even "empathize"? Drawing inspiration from frameworks like Bloom's Taxonomy (Bloom, 1956), we can envision VR experiences where students "create," "reflect," or "predict," pushing boundaries and deepening engagement.

Evolution of Reading

In the modern era, the modes of consumption have undergone a radical transformation. Recall the days when commuters on public transportation would delve into paper books, their noses buried deep in pages. Fast forward, and now the scene has morphed into individuals engrossed in smartphones and Kindles, swiping and tapping away. As discussed in an earlier chapter, today's young generation grows up engaged in multimodal interactions, with devices like iPads being commonplace from a very young age. This technological evolution has set the stage for a paradigm shift in content consumption: from reading to experiencing. Imagine diving into Tolkien's Middle Earth, not merely through words but as an active participant, engaging with its lore and characters. This immersive experience isn't exclusive to fictional realms. Consider the evolution of the Harry Potter franchise. Beyond the books and films, Warner Bros. Discovery and NEON are unveiling "Harry Potter: Visions of Magic," an immersive art experience that transports fans into iconic places in the Wizarding World, complete with responsive video content, bold architecture, original soundscapes, and multisensory installations (NEON, 2023).

For today's youngsters in particular, such digital immersion is a given; it's a world where efficiency and immediacy rule. As a compelling testament to this generational shift, a 2011 viral video showcased a one-year-old girl instinctively trying to interact with a paper magazine as if it were a touchscreen. Jean-Louis Constanza, her father, posited this as evidence of a generational transition propelled by technology (Jabr, 2013). Such observations underscore that, for many of today's youth, the tactile and noninteractive nature of traditional paper can feel unfamiliar. This shift isn't just about preference or familiarity; it has profound cognitive implications. The human brain perceives text as a tangible element in the physical realm (Jabr, 2013). Reading from paper involves specialized cognitive processes, offering a sense of spatial navigation that screens sometimes lack (Shi et al., 2020). This tangible aspect of paper books provides readers with a clear structure, enhancing memory retention and content navigation (Jabr, 2013). Conversely, digital platforms, including e-readers and tablets, might interrupt this intuitive navigation, potentially inhibiting the formation of a coherent mental representation of the text (Jabr, 2013). Research has backed these observations; some studies suggest that screens might impair comprehension, influencing the long-term retention of information (Clinton, 2019; Delgado & Salmerón, 2021; Mangen et al., 2013). This was evident in a study involving college students: When pressed for time, both digital and paper readers performed similarly (Ackerman & Goldsmith, 2011). However, with better time management, paper readers outperformed their digital counterparts by 10 percent, indicating a deeper, more focused engagement with content (Ackerman & Goldsmith, 2011).

Despite the cognitive advantages of paper, the younger generation's inclination toward screens is undeniable. They are less biased against digital media, but there's a nuanced perception at play. E-books, for instance, are sometimes seen as more ephemeral compared to their paper counterparts. Yet, there's a parallel here with the music industry. Just as initial resistance to digital music formats gradually waned, attitudes toward e-books are likely to evolve, especially as technological advancements bridge the experiential gap. Engineers are already working to make e-readers mimic the tactile feel of paper, and innovative interfaces are being designed to amplify the reading experience on tablets. Furthermore, the capabilities of screens extend beyond mere text representation. Digital platforms can offer scrolling, interactive tools, and various forms of engagement that paper simply can't replicate. This suggests that while paper might still be preferable for deep, immersive reading, the digital medium is carving out its niche, especially in a world where multimodality and interactivity are increasingly valued. These advancements hint at an impending future where our commutes might not just be accompanied by e-readers or smartphones, but perhaps, immersive metaverse experiences. Instead of simply reading about another world, travelers could potentially step into one. As technology continues its relentless march forward, intertwining ever more with the arts and daily life, the lines between reading and immersion are set to blur further, crafting a reality where narratives are not just consumed but lived.

Evolution of Gaming

Interactivity is a vital aspect of the metaverse; participants are not passive but active cocreators of the virtual world. Historically, video games began as straightforward, linear experiences. The early days of video gaming featured titles such as *Pong, Space Invaders, Tetris, and Super Mario*. Simplicity marked these games in terms of graphics, gameplay mechanics, and objectives. They catered mainly to individual players, either pitting them against the computer or facilitating competition against a local adversary. In these games, players engaged with a somewhat predictable environment. As time progressed and technological innovations advanced, coupled with the intrinsic social nature of gaming, a significant shift emerged. The focus transitioned to more expansive, open, and collaborative gameplay experiences. Modern games often emphasize player interaction, collaboration, and global competition. This shift can be attributed to the popularity and advancement of multiplayer online games that connect players worldwide. Titles like *Fortnite* and *Minecraft* showcase this shift, providing environments where players can collaborate, create, and even compete. This move toward a participatory model seems to align with broader cultural trends. Audiences nowadays don't just want to observe; they desire active, meaningful engagement.

The rise of e-sports, which mimics the structure and enthusiasm of traditional sports, supports the shift toward participation. It mirrors traditional sports, allowing fans to revel in watching professional players while also participating in amateur leagues or enjoying the game at a recreational level. The gaming industry has skillfully harnessed this participatory ethos, creating vast, vibrant communities. Platforms like Twitch empower players to stream their gameplay, enabling real-time interactions with their audience. This platform has not only redefined the notion of gaming celebrity, but it also epitomizes the collective and participatory nature of modern gaming. The gaming landscape has thus evolved dramatically. It has transitioned from individualized, solo experiences to a world defined by shared, collaborative, and deeply interconnected ventures. The compelling drive to participate, collaborate, and be an active part of a larger community stands as a defining feature of the contemporary gaming world. Joy and fun play a vital role in motivating immersion in the metaverse. The real-time interaction and immersive nature of the metaverse bring a heightened sense of enjoyment to gamers, leading to a migration from simple online games to more virtual experiences. Unlike common digital games today, the metaverse offers continuous real-time action without pauses, game overs, or resets. Additionally, with fast internet connections and powerful VR headsets, users can have immersive experiences, including 360-degree views of the digital environment.

Evolution of Entertainment

It is not only gamers whose digital experiences are being transformed, the general public's communication and consumption habits have also been revolutionized. Entertainment and communication have significantly progressed, propelled by both technological advancements and the ever-changing needs of audiences. A notable trend reshaping the entertainment canvas is transmedia storytelling. At its core, transmedia storytelling crafts content that is not tethered to one medium but fluidly transitions and adapts across multiple platforms (Jenkins, 2003). This strategy ensures content has both a wider reach and offers a more profound engagement. South Korea's Kakao stands as a testament to this transformation. Their innovative approach to integrate webtoons, novels, and dramas has positioned them as pioneers in the transmedia arena. Take, for instance, the evolution of the web novel *Business Proposal*. From its humble beginnings as written content, it seamlessly transitioned to a webtoon and eventually made its mark as a globally renowned TV series available on Netflix (Kim, 2022). This isn't just about repackaging content; it's about reimagining and expanding the narrative in unique ways, allowing audiences to engage with the story through multiple lenses.

The march of progress doesn't halt there. The rise of over-the-top (OTT) platforms, coupled with the principles of Web 3.0, has ushered in an era where content is not just diverse but also more accessible. Streaming giants like Netflix, Amazon Prime, and Disney+ harness the power of data analytics, tailoring content to resonate with individual user preferences. With the potential integration of Web 3.0 technologies, OTT platforms could champion a new wave of entertainment, fostering direct, decentralized bridges between content creators and their audiences (Kohli, 2023). This decentralization would challenge traditional hierarchies in the entertainment world, enabling creators to both distribute and monetize their work without leaning on centralized entities (Punzo, 2022). The implication here is profound: a seismic shift in power dynamics, offering more agency to creators and ensuring a more transparent, democratic entertainment ecosystem.

The boundaries between creators and consumers are becoming increasingly porous. With the combined force of OTT platforms, Web 3.0, and the transmedia phenomenon, the entertainment industry is experiencing a paradigm shift. Audiences are no longer mere spectators; they're integral to the content creation process, making entertainment more interactive, personalized, and immersive than ever before. Embracing the spirit of Web 3.0 means delving into deeper personalization, allowing users a level of control over their digital experiences that was previously unimaginable. This leads to a significant metamorphosis in user behavior. Audiences are no longer just passive spectators on the sidelines; they evolve into active participants, cocreating the content they consume. Platforms like TikTok and YouTube are at the forefront of this shift, celebrating and facilitating the democratization of content creation. Here, every user becomes a storyteller, weaving their unique narratives. Transmedia storytelling, with its adaptable nature, finds a fertile ground in such platforms, enabling tales to span across mediums—be it text, animation, or live-action. And as users get the tools and agency to remix, reshape, and expand these stories, the world of entertainment transforms into a collaborative tapestry, woven together by creators and consumers alike.

The Digital–Physical Fusion: Shopping as a Paradigm

As society evolves, so do the ways we engage with the world around us. This shift is evident in numerous sectors, but one of the most illustrative examples can be found in our shopping habits. A decade ago, shopping typically involved visiting a physical store, examining a product, perhaps reading its label, and then making a purchase. Contrast this with the contemporary shopping experience, which often begins online—browsing reviews, watching product demonstration videos, and gathering insights. Armed with this digital knowledge, today's consumers may still visit a brick-and-mortar store, but now they often simultaneously reference digital resources, such as price comparison apps or online reviews. This intricate interplay between physical and digital realms underscores a significant evolution in consumer behaviors. The metaverse exemplifies this merging of worlds. Within it, consumers might use AR to virtually "try on" a product or solicit feedback on social media before making a purchase decision. A product's online reputation can profoundly influence its real-world appeal.

Yet, such changes aren't confined to the purchase process. The way we interact with products post-purchase has also transformed. While printed manuals once reigned supreme, many now turn to online tutorials or community forums. An increasing preference for video tutorials over traditional written instructions signals a notable shift in our learning preferences. This melding of the digital and physical isn't exclusive to shopping. The metaverse stands as a testament to this convergence—a realm where our online and offline experiences not only coexist but seamlessly intertwine. Consumers turning to AR or crowdsourced opinions for shopping mirrors how the metaverse influences multiple facets of our daily lives. Today's consumers, accustomed to the immediacy of digital solutions, frequently bypass traditional resources in favor of online alternatives. The shift toward visual and interactive mediums, such as YouTube tutorials, exemplifies the depths to which our lives are enmeshed in this digital–physical blend. It's evident that the metaverse isn't just a space we visit; it's a reflection of our contemporary existence. And while this discourse focuses on the realm of shopping, similar narratives can be discerned across sectors. It's a testament to the pervasive influence of the metaverse and an indication that our ways of engaging, learning, and consuming will only continue to evolve.

Emotions in Online Spaces

While some critics argue that communication within the metaverse might lack emotion and empathy, this perspective doesn't consider the entirety of human experience. It's essential to avoid a black-and-white view of this issue. Individual differences play a significant role in how we communicate and interact with the world around us. Not everyone thrives or feels most expressive in physical environments, and for some, the digital realm might be where they find their voice. The metaverse, with its evolving technologies, is designed to foster efficiency, expressivity, and even empathy. Just as face-to-face interactions vary in depth and quality, so too can interactions in the metaverse. Given these individual differences, it's crucial to recognize that over time, as the technology matures and as users become more adept, the emotional richness of metaverse interactions can match—if not exceed—traditional modes of communication.

In a study analyzing online interactions during the peak of social distancing, Tibbetts et al. (2021) discovered that there were emotional differences between online and offline interactions. It was found that in-person interaction with people to whom one is close is likely to generate more social connection. Online interaction with such people also results in greater social connection as well. However, on the flipside, online interaction with people that one is less close to is likely to create more negative feeling and stress. Lieberman and Schroeder (2020) identified four main differences between online and offline interaction—namely, nonverbal cues, anonymity, development of social ties and information circulation—and how online communication can either interfere with or improve offline communication. Okdie et al.'s (2011) study on judgments made by interlocutors about each other during virtual versus in-person communication found that participants were more likely to give positive opinions about their fellow interlocutors and have a greater level of consensus in their judgments of one another during in-person communication as compared to virtual communication. In the context of teacher–student communication, Wiyono et al. found that offline interaction between teachers and students is more productive than its online counterpart. Laghi et al.'s (2013) study of online and offline communication among children with and without shy dispositions discovered that those with a shy personality tended to use online modes to communicate negative emotions and negative experiences with their peers, compared to those without a shy personality, suggesting that the mode of communication could be related to loneliness. Differences between online and offline communication could not be clearly discerned in certain studies. For example, Lee et al. (2011), in their investigation of the impact of in-person versus virtual communication on quality of life, found that while the former could predict quality of life, the latter failed to do so.

Exploring Emotions in the Metaverse: A Dual Perspective

Playing in the metaverse represents a new frontier of human interaction, bridging traditional feelings of joy, excitement, and anticipation with the technological advancements of AR. By enhancing our surroundings with virtual elements, the ordinary becomes extraordinary, turning mundane environments into thrilling playgrounds. The creativity and exploration that come with traditional play are further stimulated in the metaverse as AR enables players to augment the real world with fantasy elements, subtitles, emoticons, and emojis. This multimodal content, engaging our senses in new ways, offers novel paths for discovery and innovation. Social connections within the metaverse transcend physical boundaries, allowing for shared experiences that foster deeper friendships. Perceiving emotions through avatars and emojis adds a new layer of intimacy, while the metaverse offers an escape into fantastical worlds for relaxation and stress relief. Yet it's not merely about escapism; AR layers additional meaning onto our real world, creating new avenues for enjoyment and relaxation. However, this immersive experience can also lead to frustration, confusion, and sensory fatigue. The interplay of real and virtual elements can be overwhelming, especially if not well designed or engaged with excessively. Despite these challenges, mastering the virtual landscape can be deeply empowering, as AR provides tangible feedback that resonates both virtually and physically. The blending of reality and fantasy amplifies feelings of achievement, and the metaverse's shared space fosters a sense of community and inclusivity. As we continue to augment reality with virtual elements, we must also consider the ethical implications. Ensuring that this blending of reality and fantasy remains respectful of individual rights, privacy, and well-being is paramount.

Playing in the metaverse offers a profound emotional experience, taking traditional feelings associated with play and enriching them with additional layers of meaning, sensory stimulation, and complexity. The metaverse challenges us to navigate a new landscape with curiosity, empathy, and responsible stewardship, as it brings together the real and the virtual, the mundane and the extraordinary, in ways that were previously unimaginable. Immersive VR transports you to a world where your dreams and imagination come alive. When you wear a VR headset, you enter a realm of your own creation, free from distractions in the real world. Unlike flat screens, which can constantly draw your attention away, VR provides a more engaging and immersive experience; once you put on the headset, your experience becomes isolated and hidden from the view of others in the physical world. It's like stepping into a personalized reality tailored to fulfill your desires and aspirations. The VR headset acts as a gateway, enabling enhanced focus compared to traditional 2D online screens. Within the immersive metaverse, you have the freedom to explore and fully immerse yourself in your VR adventure.

Empathy Matters

The rise of VR and metaverse platforms has ushered in innovative avenues for developing empathy. Research evidence suggests that when users step into the shoes of an avatar in VR, it can pave the way for transformative experiences that enrich human connection. For instance, users have exhibited a reduction in implicit racial biases after embodying avatars of different ethnicities (Peck et al., 2013). Similarly, VR has the potential to instill empathy for unique challenges, like understanding the worldview of someone with color blindness (Ahn et al., 2013). Moreover, these digital spaces cultivate prosocial behaviors, as evidenced by users' responses when confronted with scenarios wherein virtual beings require assistance (Ahn et al., 2013; Gillath et al., 2008). While these outcomes offer promise, it's essential to recognize the developmental intricacies, especially among younger populations. For young children, who are still nurturing their capacity for perspective-taking and refining their understanding that others might perceive the world differently (Blakemore & Mills, 2014), VR can be a complex terrain. However, as they evolve into adolescence, these virtual experiences could further fine-tune their empathetic abilities. Considering today's Gen Z, their empathy and sense of solidarity are notably interwoven with their online experiences. For many in this generation, social media isn't just a platform for communication but a tool for understanding, expressing, and fostering empathy. They engage in global conversations, share experiences across boundaries, and rally behind causes with fervor. In such a landscape, the metaverse isn't a mere extension but a profound tool that can amplify this digitally driven empathetic surge, offering a more immersive and interconnected understanding of diverse human experiences.

Multisensory Communication in the Metaverse

The concept of a multisensory experience holds paramount importance in the realm of immersive encounters. This notion underscores the significance of engaging multiple senses simultaneously to create a profound and captivating journey. In an immersive experience, the human senses are no longer isolated, but rather harmoniously orchestrated to form a cohesive symphony of perception. Incorporating visual elements is just the beginning; a multisensory approach extends to encompass auditory, tactile, and even olfactory sensations, intertwining them to craft an intricate tapestry of perception. Visual cues provide the foundation, shaping the environment and setting the stage for the experience. However, it is the convergence of auditory stimuli that transports participants further into the narrative. A carefully composed soundtrack, nuanced with varying tones and volumes, guides emotional responses and enhances the sense of presence. The tactile dimension, often overlooked, plays a pivotal role in anchoring participants within the experience. The sensation of touch can range from the subtle vibration of a controller to the more complex haptic feedback that replicates physical interactions. This tactile feedback intertwines with the visuals and sounds, fortifying the immersion by grounding it in the realm of the tangible. Additionally, the sense of smell can introduce an unexpected layer of authenticity. The ability to recreate scents associated with a particular environment or moment can evoke deep emotional responses, enriching the overall multisensory experience. By capitalizing on the synergy of multiple senses, immersive experiences transcend the limitations of any single medium. A truly immersive encounter captivates not only the mind's eye but also the ears, the fingertips, and even the nose. This cohesive blend of sensations crafts a reality that feels vivid and tangible, resonating deeply with our perceptual instincts and evoking emotions that are indelibly etched into memory. As technology advances and creative boundaries expand, the role of the multisensory experience remains an essential cornerstone in forging connections between the virtual and the real.

Multisensory communication in the metaverse is moving fast. While it's not quite like face-to-face interaction, it can serve as a good bridge. How closely it resembles real physical interaction depends on technology. Senses play a crucial role in face-to-face communication, and this remains true in metaverse communication. However, a challenge arises when it comes to replicating the full range of human senses in artificial spaces. Humans have five primary senses: sight, hearing, taste, smell, and touch. Through our sense of sight, we perceive the world around us, recognizing shapes, colors, and depth. Our sense of hearing allows us to listen to sounds, speech, and music, enabling communication and understanding. The sense of taste allows us to differentiate between flavors, savoring the sweetness, saltiness, bitterness, or sourness of foods and drinks. Our sense of smell enables us to detect various scents, from fragrant flowers to delicious foods. Our sense of touch allows us to feel textures, temperatures, and pressure, providing us with a tangible connection to the physical world. While incorporating these senses into artificial spaces, especially beyond vision, remains a formidable challenge, advancements in haptic and audio technologies hold promise in simulating touch and delivering immersive auditory experiences. The ongoing progress in these areas brings us closer to a future where the richness of human senses can be experienced within the virtual realms.

Vision Matters

Visual senses are highly advanced in the metaverse. One of the cutting-edge technologies in this realm is holography. In 1951, Dennis Gabor made a significant discovery in the field of visual imaging by finding a method to create images that appeared three-dimensional. Gabor's approach relied on the principles of interference and coherence, which involve the interaction and alignment of light waves (Nobel Prize, n.d.). The process involved capturing light from an object on photographic film alongside a reference beam that did not interact with the object (Nobel Prize, n.d.). This ground-breaking development marked the inception of holography, a revolutionary technique in visual representation. Holograms enable the display of high-quality, three-dimensional digital representations of individuals without the need for viewers to wear a headset. The creation of a hologram involves manipulating light beams through interference, capturing reflected light from real physical objects and people, and recreating visually stunning representations with remarkable depth and realism.

Holography has emerged as an empowering tool across various fields, including medicine. In the medical domain, holograms provide innovative means of visualizing intricate anatomical structures and medical data. By projecting holographic representations of organs, bones, or patient scans, doctors can deepen their understanding of a patient's condition and make informed decisions regarding diagnosis and treatment options. The use of holograms in medicine has the potential to revolutionize medical education and surgical procedures. Medical students can benefit from immersive and interactive holographic simulations, enabling them to practice procedures and explore anatomy in a more tangible and realistic manner. Surgeons can leverage holographic overlays during operations to enhance precision and minimise risks by superimposing vital information directly onto the patient's body. Holograms serve as a powerful tool for doctors, enhancing their ability to visualize complex medical information and improve decision-making processes. As technology continues to advance, holography evolves, promising even greater advancements in the medical field and beyond. These advancements hold potential in diverse applications, expanding the possibilities for enhancing education, training, and patient care.

Sound Matters

Sound and audio not only enhance engagement but also convey intelligence and significance in communication. High-quality audio creates an impression that your words carry weight and that interacting with you is a pleasurable experience. A recent study has revealed that good sound fosters a sense of affinity (Newman & Schwarz, 2018). Sound and audio enrich the user experience by adding emotions and nuances to the input, making it more powerful, assertive, or delicate. They can simulate depth, replicate sounds, and enhance enjoyment. However, sound and audio are still in the early stages of development, facing challenges such as the need for high-quality hardware and efficient streaming. Nonetheless, their immense potential is evident. It is important to address technical obstacles and ethical considerations, especially in terms of using someone else's voice from a library, as it raises questions of appropriateness and authenticity. Despite these challenges, sound and audio offer vast possibilities for shaping an exciting metaverse when coupled with the right technology and approach. There is still much to explore regarding our perception of sounds and their implications within the realm of VR. Further research is necessary to unravel the intricacies of sound matters in the metaverse. Additionally, text-to-voice chat and proximity chat are emerging features being added to enhance communication in the metaverse. These technologies aim to provide more immersive and realistic interactions by allowing users to convert text messages into spoken words and engage in voice conversations based on proximity in the virtual environment. However, it is important to note that these features are still in the early stages of development and require further advancements to fully meet user expectations and provide seamless and high-quality experiences.

Touch Matters

Haptic communication has the potential to transform our interactions and experiences. It allows us to perceive and feel through machines, creating a more immersive and engaging connection with our digital surroundings. Imagine being able to touch and feel virtual objects, textures, and sensations as if they were real. Haptic feedback adds a new dimension to VR and AR experiences, making them more realistic and believable. It can also enhance our gaming adventures, telepresence interactions, and various other applications.

However, there are still some challenges to overcome in the realm of haptic communication. Achieving realistic and natural haptic feedback can be complex, requiring advancements in hardware and software development. Additionally, implementing haptic technology can be costly, making it a consideration for widespread adoption. Despite these challenges, the field of haptic communication is evolving, and its potential impact is significant. As technology continues to advance, we can expect haptic communication to find its way into more aspects of our lives. Imagine feeling the impact of a virtual punch during a game, receiving tactile cues to improve driving safety, or even undergoing surgical training with realistic haptic simulations. The applications are vast, and the possibilities are only limited by our imagination.

Scent Matters

Scent technology is advancing rapidly. This was showcased by metaverse-focused companies at the Consumer Electronics Show (CES) 2023. Notably, luxury fragrance brand Byredo has joined forces with RTFKT Studios, NFT makers acquired by Nike, to introduce a series of digital perfumes called Alphameta that will be featured on the blockchain (Carrillo, 2022). This innovative collaboration, which draws inspiration from video game potions, allows users to equip their RTFKT Clone X avatars with the aroma of virtual perfume (Carrillo, 2022). A wide array of ingredients can be combined to create a scent, each to evoke specific feelings such as acuity, harmony, naivety, and virtue (McDowell, 2022). Additionally, The Scents of Wood Company is launching the world's first fragrance subscription, exclusively available through the purchase of an NFT (Scents of Wood, 2022).

Limitations to Current Immersive Communication: Affordability and Reliability

VR and the broader metaverse concept are undeniably intriguing in their potential for reshaping how we interact and experience the world. Yet, several barriers are currently slowing their mainstream adoption. Chief among these are the issues of affordability and technical reliability. At present, the cost of immersing oneself in the metaverse through VR is considerable. High-quality headsets and the necessary accompanying hardware can be financially prohibitive for many. Even though the increasing demand for these tools might eventually lead to reduced prices through economies of scale, there remains an immediate cost concern for potential users. Mobile apps are making strides in providing a more cost-effective entry point into VR, turning smartphones into basic VR platforms. The rise of AR and MR also shows promise for offering more affordable alternatives. For instance, innovations like the holographic communication app announced by Slovakian software company MATSUKO, which uses smartphones and integrates AR with AI (Melnick, 2022), highlight potential pathways to more accessible immersive experiences.

However, beyond the financial considerations, the reliability of these technologies is also a point of contention. Technical issues, from latency in rendering to other glitches, can detract from the seamless immersive experience these platforms aim to provide. Such disruptions, although expected in newer technologies, underscore the present limitations. The recent shift toward digital and virtual platforms, hastened by global events, brings with it a clear realization: While virtual modes offer certain conveniences, they cannot entirely replace physical interactions. Determining the right balance between virtual and physical realms remains a challenge. While the potential of the metaverse and VR is vast, the journey toward their widespread acceptance and integration hinges on overcoming present limitations, particularly in terms of cost and consistent reliability.

5

Metaverse Psychology

Virtual reality can offer a blissful experience for individuals who deal with various forms of anxiety, such as public speaking anxiety (PSA) or foreign language anxiety (FLA). In educational settings, VR provides a safe space for students to thrive, benefiting introverted individuals who may struggle with social interaction. However, there are complexities and concerns within virtual spaces that need to be explored. The research discussed in this chapter reveals the mixed effects of VR usage, highlighting both its potential as a safe and enabling space and an unsafe space.

Social Anxiety: A Complex Matter

Classroom and social anxiety are prevalent issues among students. The UK Student Behaviour Report, commissioned by Chegg's Center for Digital Learning in partnership with Hanover Research and with input from Universities UK, sheds light on the state of student mental health. According to the report, nearly three-quarters (71 percent) of students experience anxiety regarding their classes and schoolwork, while 44 percent struggle with meeting new people (Chegg, 2023). Although this data is specific to the UK, similar challenges can be observed in other regions, particularly in Asia, where hierarchical structures may exacerbate these anxieties. Sometimes, anxiety is easy to identify—like when a child is feeling nervous before a test at school—but other times, anxiety in the classroom can look like something else entirely different: an upset stomach, disruptive or angry behavior, or even a learning disorder. There are many different kinds of anxiety, and this is one of the reasons it can be hard to detect in the classroom. Social interaction doesn't come easily for everyone, and for some, social anxiety and the stress of public exposure are overwhelming. Individuals who speak a language other than their mother tongue may also face heightened anxiety. In my previous study focusing on young children and language anxiety, I discovered that severe stress manifested in physical symptoms, including hair loss, selective mutism, and depression (Kiaer et al., 2021). This led me to question whether virtual spaces could offer a solution. Specifically, I wondered if representing oneself as an avatar in the virtual world could decrease anxiety levels. My initial findings indicate that social anxiety in physical spaces is not necessarily replicated in virtual spaces.

Dilemma in Metaverse Research: Speed Matters

The metaverse encompasses both empowering and potentially endangering aspects. While it has long been associated with gaming, it has now expanded to become a phenomenon accessible to the general public, offering immersive experiences beyond gaming. Virtual spaces can be transformed into classrooms, offices, medical consultation rooms, and more. As our exploration of the metaverse continues, it becomes crucial to understand its impact. However, academic studies published as journal papers or monographs struggle to keep up with the rapid pace of technological advancements, making it challenging to fully comprehend the metaverse's potential.

The purpose of this chapter, and the book as a whole, is to delve into whether VR can serve as an empowering space, particularly for individuals with social anxiety, by reducing unease and providing them a platform for self-expression. After extensively reviewing the literature in this field, I have concluded that the answer is not so straightforward. The effects of VR on anxiety are complex, influenced by individual differences and the specific VR environment. Moreover, the constant evolution of VR environments poses challenges in applying findings from older studies to the current digital landscape. Most studies have limited generalizability due to their reliance on small and specific sample sizes, preventing translation of their findings to broader populations and situations.

The rapid development of technology leaves little time for comprehensive investigations, resulting in research outputs becoming outdated quickly. This can be discouraging for those working in the fields of AI and VR research, as there is a great need, more than ever, for insights that can benefit the general public, policymakers, and other stakeholders. Research culture has not fully adapted to the fast-paced digital landscape that we are in, making it challenging to gather the necessary insights and provide relevant and timely information.

The Challenge of Generalization in Virtual Realms

Everyone is unique, which poses challenges when attempting to generalize social interactions based solely on demographic information. Even if individuals share characteristics like ethnicity, gender, mother tongue, or age, their perceptions and behaviors in the virtual world can vary greatly. Some can effortlessly form friendships in the virtual realm, while others struggle. It's important to note that our behavior in the virtual world does not necessarily reflect our real-life experiences. The virtual environment operates according to its own set of rules, highlighting the need for caution when applying research findings to individuals. Additionally, the specific virtual environment, platform, and interactions within virtual spaces have a profound impact on one's virtual behaviors and perceptions.

In the following section, I will present several research studies that contribute valuable insights. It is important to note that most of these studies were conducted during the pre-pandemic period, which was an inevitable consequence of the significant technological advancements that occurred during that time. The literature review included in this chapter could be better categorized as pre-ChatGPT era, as it was conducted before the emergence of generative AI like ChatGPT. Considering the rapid pace of academic publications, it is likely that some of the most recent findings and discussions related to the impact of ChatGPT in the metaverse may not be included in this book. However, it is worth mentioning that the studies presented here have their own limitations, such as small sample sizes and region-specific focus, which may restrict their generalizability. Nonetheless, these studies provide valuable insights into the current state of our digital landscape and the potential impact of VR on our psychological well-being.

Metaverse: A Tool for Oral Presentations

Davis et al. (2020) conducted a study examining the potential of VR in aiding college students to practice oral communication and public speaking skills, particularly in relation to communication apprehension (CA). CA encompasses two types of speaking anxieties: situational anxiety, which is context-specific, and trait anxiety, which persists across all contexts. The study explored various strategies to alleviate CA, including exposure therapy, cognitive modification to replace negative associations with public speaking, and skills training to enhance competency. They divided participants into two groups: one group practiced delivering a presentation in a virtual environment simulating a classroom, while the other group practiced in-person with a peer. Interestingly, no significant difference was found between the two groups in terms of their results. This suggests the potential usefulness of VR as a tool for presentation practice. The study also highlighted the effectiveness of VR as a form of exposure therapy, as the virtual audience felt real to the students.

Damio and Ibrahim (2019) conducted a study to explore the potential of VR speaking applications in helping individuals overcome the fear of giving presentations. The study focused on TESL postgraduate students at UiTM Puncak Alam in Malaysia and data was collected through questionnaires and interviews. Participants expressed interest and motivation in using VR for presentation preparation, finding the experience enjoyable and engaging. However, they acknowledged that VR was not the most effective method for developing speaking skills. It is important to note that these findings are based on a single study, and results may vary depending on participants' awareness of VR and digital media literacy. Other studies may have different expectations. This highlights the need for individual instructors to assess and tailor the use of VR in the classroom, as research findings on its effectiveness are mixed.

AR: Improving Engagement and Audience Feedback

Parmar and Bickmore (2020) conducted a study where they developed an AR system to provide real-time feedback during presentations. The study revealed several key findings. First, presenters who used 3D spatial overlay visualization engaged more with audience members by addressing them by name, resulting in higher audience engagement. Second, the use of a head-mounted AR display did not have a negative impact on speaker confidence and anxiety. Lastly, judges perceived higher audience engagement when presenters used 3D spatial overlay visualization. Additionally, presenters using peripheral visualization demonstrated more appropriate use of stage space and body posture. These findings highlight the potential of AR systems, especially those incorporating 3D spatial overlay visualization in improving audience interaction and enhancing presentation skills. Overall, the study suggests that AR has the ability to improve engagement and provide valuable feedback from the audience.

The Impact of VR on Reducing Anxiety

Poeschl (2017) presents the QUEST-VR framework, which evaluates the effectiveness of VR training for public speaking. The study showed that while VR has the potential to be used in public speaking training, it does not always reduce anxiety. In contrast, Takac et al. (2019) showed how public speaking anxiety decreases through repeated VR training. They measured anxiety through physiological measurements (such as heart rate), Personal Report of Confidence as a Speaker (for measuring fear of public speaking) and subjective units of distress scale (SUDs, for measuring level of anxiety).

Boeldt et al. (2019) explored the effects of virtual reality exposure therapy (VRET), which is a type of therapy that uses VR to help people overcome anxiety disorders, such as public speaking anxiety. In VRET, individuals are exposed to virtual environments that simulate the situations they fear, allowing them to gradually confront and desensitize their phobias in a safe and controlled way. Research has shown that VRET can effectively reduce public speaking anxiety symptoms and improve quality of life. VR exposure therapy offers several benefits as compared to traditional exposure therapy. It provides a safe and controlled space for individuals to face their fears, and the high level of immersion in VR makes the experience feel more realistic, enhancing the learning of coping skills. There are several factors to consider when it comes to VR exposure therapy. First, the cost of this therapy can be high due to the specialized equipment needed for the treatment. Additionally, VR exposure therapy is not yet widely accessible as there are only a limited number of therapists who are trained in providing this type of therapy. Another obstacle is that VR technology is constantly advancing, so newer versions of VR headsets will offer improved effectiveness compared to older versions, but this comes at a cost. Lastly, individuals may have different responses to VR exposure therapy. While some people may find it highly effective, others may not experience the same level of benefit.

Systematic Exposure and Individual Differences

Yadav et al. (2019) investigated the association between bio-behavioral indices and public speaking anxiety, as well as the effectiveness of VR exposure therapy in reducing PSA. Participants delivered presentations in VR environments and completed questionnaires on individual and contextual factors. The study revealed the variability in participants' experiences of PSA and emphasized the importance of considering individual and contextual factors when estimating PSA. Additionally, systematic exposure to public speaking in VR was found to effectively reduce PSA based on self-reports and bio-behavioral indices.

In a similar study, von Ebers et al. focused on using wearable device data to measure the effectiveness of VR exposure therapy in reducing PSA. They conducted an experiment with twenty-three participants who received public speaking training in VR environments while wearing devices that collected physiological and acoustic data. The findings showed that bio-behavioral data, particularly acoustic indices, could predict the effectiveness of VR in treating PSA. Demographic factors such as gender, age, and accent significantly influenced the acoustic indices. Both acoustic and physiological indices demonstrated improvements in PSA, and reliable changes in PSA were observed after six to eight VR training sessions.

Exploring the Impact of Virtual Bystanders: A Study on Public Speaking Performance

Just like real-life communication, VR communication is influenced by multiple factors, such as whether you are alone, with familiar individuals, or with strangers. It is important to note that one's behavior in real life may not necessarily reflect one's behavior in virtual environments. Qu et al. (2015) investigated the influence of virtual bystanders (a virtual audience) on students' public speaking performance in language lessons. The study aimed to understand the effects of virtual bystanders' beliefs, attitudes, and behavior on students' self-efficacy, anxiety, and engagement. The study involved 26 non-native English-speaking university students, and various measurements, such as self-report questionnaires and physiological indicators, were recorded. The findings indicated that virtual bystanders with positive attitudes positively influenced participants' self-beliefs, engagement, and reduced anxiety. However, the impact of inconsistent bystander attitudes and beliefs on participants' beliefs was not clearly concluded. Participants also experienced increased anxiety when virtual bystanders showed negative attitudes toward other virtual speakers. Overall, the study demonstrated that virtual bystanders can significantly influence students' experiences in virtual environments, similar to real-life situations.

VR and Foreign Language Anxiety

Foreign language anxiety (FLA) is becoming a critical issue as our world becomes increasingly multilingual. Specifically, English has become a significant source of anxiety for many foreign students studying in English-speaking higher education institutions. Foreign language anxiety is not limited to students alone. People living or working in English and other languages also experience this anxiety, and as the world becomes more multilingual, the stress associated with it increases. This is especially true because English is widely spoken as the lingua franca, putting pressure on individuals to speak English proficiently. This anxiety affects people of all ages, from young to old.

Fondo and Jacobetty (2020) proposed the Telecollaborative FLA Scale (T-FLAS) to assess and understand the presence and impact of FLA during virtual exchanges. The study involved two projects: a one-to-one bilingual exchange between Spanish-speaking students and English-speaking students majoring in Spanish, and a one-to-many monolingual exchange between Spanish and English speakers. The T-FLAS framework, adapted from Horwitz et al.'s (1986) Foreign Language Classroom Anxiety Scales (FLCAS), was administered as a survey at the end of the experiment. The findings revealed that anxiety in virtual exchange contexts can be categorized into foreign expression and learning as well as virtual environment components. Four types of students were identified: apprehensive communicators, anxious learners, technophobes, and confident communicators and learners. Interacting online was generally not problematic except with strangers, and anxiety primarily stemmed from social interaction and self-consciousness when using the foreign language. Technical issues remained a challenge for the students.

Godefridi et al. (2021) explored the impact of VR on reducing public speaking anxiety in L2 English learners. The study had a small sample size, limiting the generalizability of the findings. Nine participants underwent VR training and received peer feedback to enhance their presentation skills. Surveys, including Davis's Technology Acceptance Model and McCroskey's Personal Report of Public Speaking Anxiety, were used to measure participants' perceptions and anxiety levels. The results showed a decrease in pre-presentation anxiety after the VR training. Participants found the VR training beneficial for improving their presentation skills and managing stress. Peer feedback was perceived as less stressful than teacher feedback.

Kaplan-Rakowski and Gruber (2023) investigated the impact of VR on reducing foreign language anxiety during public speaking practice. The experiment involved twenty college students who were split into a control group using Zoom and an experimental group using VR. Participants discussed eight topics and completed pre- and post-test questionnaires on demographic details and anxiety. The results showed that practicing public speaking in VR significantly reduced foreign language anxiety compared to practicing through video conferencing. The study suggests that increased and repeated exposure to VR could further enhance the effectiveness of lowering foreign language anxiety.

Lear (2020) explores the potential of VR exposure therapy for reducing speaking anxiety in the L2 classroom. The study reviews literature on anxiety in language learning and discusses the technical capabilities of VR and its usage in exposure therapy. Lear also examines learner attitudes toward VR, addresses implementation in classrooms, and highlights disadvantages such as visual discomfort and learning curve. While advocating for the use of VR exposure therapy, the study emphasizes the need for further research to confirm its effectiveness.

As highlighted earlier, it is crucial to translate these findings into practical applications for policymakers, educators, and the general public. It is important to note that meaningful research in this field has only recently started emerging in the past few years. The limitations of previous studies arise from the rapid evolution of technology, resulting in varied individual experiences with VR. Moreover, the qualitative studies conducted thus far have focused on specific demographics and small sample sizes, limiting their direct applicability. To address these limitations, further research is necessary to gain a comprehensive understanding. However, conducting and disseminating research in the decentralized landscape of the Web 3.0 era poses significant challenges. Moreover, the cost associated with VR raises concerns about its cost-effectiveness in relation to the achieved outcomes, further warranting careful consideration.

Lyu (2019) found that the AR application *GOAT*, with its gamification features, helped Japanese English as an additional language (EAL) learners overcome language anxiety. Participants reported increased confidence and motivation to speak English, as well as a more enjoyable learning experience compared to traditional methods. However, the exact impact of AR on anxiety reduction and confidence building remains uncertain.

Reduced Anxiety, Increased Enjoyment

Satake et al. (2021) explored the effects of using VR in English conversation classes for Japanese students. The study revealed reduced anxiety and increased enjoyment among participants, indicating the effectiveness of VR in facilitating practical English conversation skills. These findings emphasize the positive impact of VR on learners' attitudes and language learning experiences.

Similarly, Thrasher (2022) investigated the effects of VR on twenty-five L2 French learners. Participants exposed to group tasks in a VR environment reported lower anxiety and demonstrated reduced physiological stress (measured by salivary cortisol). Their speech in VR tasks was also rated as more comprehensible as compared to traditional classroom tasks. The findings suggest that VR experience and increased speaking practice contribute to anxiety reduction and improved oral production in language learning.

York et al. (2021) investigated the influence of voice, video, and VR communication on the foreign language anxiety of EFL learners and their language learning experiences. The study involved thirty Japanese university students who completed spot-the-difference tasks and pre- and post-test questionnaires. The results revealed that all three modes of communication effectively reduced foreign language anxiety, with no significant differences between the modes. However, participants considered VR to be the most enjoyable and effective learning environment. Participants' individual technology dispositions may have influenced their anxiety levels and perceptions of the different modes. Additionally, the use of avatars for anonymity in VR did not impact participants' overall perception of the modes.

Individual Difference Matters

During a one-week digital humanities workshop that I conducted in February 2022, many students shared positive feedback about their experiences with VR learning platforms enhanced by AI, such as ChatGPT. I consistently observed the heightened engagement and interactivity among participants. They expressed that learning became more enjoyable in such environments compared to traditional methods. A prominent sentiment was the liberty to pose questions, even those they'd usually consider too trivial or "silly." This freedom to ask "silly" questions without judgment seemed to alleviate anxiety, creating a more open and curious learning atmosphere.

However, while these observations present a promising avenue for AI-enhanced education, it's crucial to recognize the vast individual differences that exist among learners. While the overall trend suggests increased engagement and reduced anxiety, we cannot blanketly apply these findings to all. Some participants showcased what can be termed as "new tech anxiety," a reservation or apprehension toward embracing novel technological interfaces or methods. This is a testament to the inherent variability in how individuals interact with and perceive new technologies. It's easy to fall into the trap of generalizing based on overarching trends, but the diverse responses underscore the importance of personalizing educational approaches. No matter how advanced or refined AI becomes, the individuality of human learners remains paramount.

Improved Learning Output, Motivation, and Reduced Anxiety

Wang et al. (2021) found that incorporating VR with visual prompt scaffolding (VPS-VR) led to improvements in EFL learners' reading comprehension skills and learning motivation. The study involved 98 students at a Chinese university. The VPS-VR group exhibited higher proficiency and reported more positive learning experiences compared to students in traditional classroom instruction or VR without VPS. Additionally, the VPS-VR approach reduced learning anxiety associated with EFL reading comprehension. Specifically, VPS-VR-trained students demonstrated enhanced text comprehension skills, which are essential components of reading comprehension. Still, one must consider the contextual factors when implementing this approach. The study was conducted at a Chinese university, known for its technological advancements. The effectiveness of VPS-VR may vary in different countries and contexts. Therefore, assessing participants' digital and AI literacy, experience, awareness, and the digital environment they are accustomed to becomes increasingly important in research.

Avatar Influence: How Virtual Bodies Shape Our Minds

An interesting phenomenon often occurs in immersive virtual reality: People start behaving differently after assuming a virtual body that reflects their real-life movements but differs in other characteristics, as if it were their own. In a fascinating study by Banakou et al. (2018), it has been discovered that virtually being Einstein has surprising effects on cognitive task performance and age bias reduction. Participants who embodied an Einstein-like virtual body exhibited improved performance on cognitive tasks compared to those who embodied a normal virtual body, even after accounting for their prior cognitive ability.

This improvement was particularly significant among individuals with low self-esteem. Furthermore, embodying the Einstein virtual body resulted in a decrease in implicit bias against older people. These findings underscore the potential of virtual body ownership to enhance cognitive functioning and foster more positive attitudes toward aging. The ability to assume different virtual identities opens up exciting possibilities for utilizing virtual reality as a tool for cognitive enhancement and promoting inclusivity.

Ethical Concerns

Yet one notable aspect that has become evident with the increasing traffic between realities and the constant "transversing" of the metaverse is the emergence of ethical issues that demand urgent attention and resolution. These ethical concerns can contribute to making the metaverse a somewhat daunting and unsettling space. While our lives operate within the finite system of being born and eventually dying, the virtual world often experiments with blurring the boundaries of life and death. Even the concept of virtual offspring is becoming a reality. Virtual offspring could potentially gain full acceptance in society, require minimal cost for upbringing, and even bypass the need to grow up.

People are not only engaging in relationships with virtual humans but are even marrying them. An interesting example is Gatebox, a Japanese company that has developed 3D AI holographic characters residing in a glass jar. These virtual companions can perform various tasks, such as reading the news, playing music, providing weather updates, and controlling household appliances (Caldwell, 2021). They go beyond being mere technological devices; around 4,000 individuals have taken their digital partners as spouses in Japan (Caldwell, 2021).

This kind of boundary-crossing raises important ethical considerations that require a nuanced perspective. In a South Korean documentary, a grieving mother was emotionally reunited with her deceased daughter through VR. Wearing VR goggles, a mother tearfully witnessed her seven-year-old daughter reappear in a neighborhood park, where she used to play before succumbing to a blood-related disease three years earlier (Park, 2020). The emotional power that VR can hold over us raises questions about our societal and ethical views on the use of VR in confronting our fundamental realities.

Artist Shaun Gladwell has created an immersive near-death experience that guides people through the stages of life de-escalation, from cardiac failure to brain death (NGV Melbourne, n.d.). This provides individuals with a glimpse into what their final moments might entail. These cases have a profound impact on our society that cannot be ignored. They require serious engagement and thoughtful consideration of the implications they hold. It is important for us to actively address these emerging technologies and their effects rather than turning a blind eye.

Investigating the Psyche of the Metaverse: Diverse Journeys in VR

As our digital landscape rapidly evolves, the metaverse and VR emerge as fore frontiers, offering tantalizing glimpses into the future of human interaction, entertainment, and learning. With a desire to delve deeper into this new frontier, I embarked on an in-depth exploration, interviewing a diverse set of VR users, each bringing their own unique perspectives and experiences to the table. What became evident early on was the complexity of unravelling the psychology of the metaverse and its users. Generalizing their experiences is not only challenging but perhaps even misguided. Like a prism of light, each user reflects different shades based on their needs, aspirations, and encounters within the VR realm. With this backdrop in mind, we now present four in-depth interviews, spotlighting the individual stories, motivations, and reflections of our interviewees as they navigate the vast and ever-expanding universe of VR.

Entering the World of VR: Insights from a Player's Experience

When exploring the dynamic world of VR gaming, it's essential to understand the experiences of those at its heart: the gamers. In a recent interaction with a VR enthusiast who I will call "Amy" as a pseudonym, we delved deep into their journey from discovery to current engagement. Beginning with an origin story, Amy reminisced, "I started using VR when I was around 18 years old. It was bought as a present for the whole family." From initial infatuation, spending "almost every day for a few hours" on it to a gradual diversification of interests, their VR trajectory is a relatable one. Amy elaborated, "I mostly used it for free-to-play games, then I began to play rhythmic games like Beat Saber." Blending passion with academia, Amy found innovative ways to incorporate VR into their studies. "I studied graphic design at secondary school," Amy shared, "so I used VR to design logos, draw different figures, portraits, and so on."

While many immerse themselves in the virtual community, Amy's interaction has remained refreshingly tangible. "I normally don't interact with other users through online platforms or chat rooms," they explained, "but I used to participate in local competitions. I only interacted face-to-face with other gamers." Discussing the merits and pitfalls of VR gaming, Amy weighed in with an optimistic but cautious perspective. They expressed, "VR is the future of gaming as it is a unique way of interaction. I quite enjoy the anonymity of online shared spaces with their colourful surroundings." Yet, they didn't shy away from addressing the challenges, noting that "VR's disadvantages include heavy headsets, malfunctioning body tracking, and glitches that can happen in the game for various reasons."

Perhaps the most poignant part of our conversation revolved around the impact of VR on one's health. Amy revealed some personal struggles, stating, "I have personally experienced a lot of physical health problems while playing in VR. As a person that suffers from epilepsy, playing in VR can be dangerous if I lose track of time. I have experienced a lot of migraines because of extensive gaming." Offering a solution, Amy suggested the introduction of a "time limit feature" in games to safeguard gamers' well-being.

Navigating Virtual Socials: Insights from a University Student

VR has found its way into various aspects of our lives, and its influence on today's youth, especially university students, is undeniable. In a recent interaction with a person who I will call "Bill" as a pseudonym, a first-year Master's student aged twenty-five, we navigated the depths of their VR engagement, from passionate gaming to the realm of virtual socials. Reflecting on their VR initiation, Bill recollected, "I used to play a lot of multiplayer games online. When I was in high school, I played every day after school." However, like many young adults juggling studies and work, Bill's current gaming schedule is more limited, restricted to "mostly during the weekends."

A shift in Bill's VR interests emerged with the discovery of virtual social spaces. Bill highlighted, "I discovered VR chat clubs and parties around 1 or 2 years ago." This wasn't just about gaming anymore, but the allure of a more personal connection. "In VR chat clubs and parties, you can meet people from all around the globe, and it feels more personal than regular chatting via websites like Facebook, Instagram, and so on." While many are cautious about online interactions, Bill's journey in the virtual domain is predominantly populated with unfamiliar faces. "Yes, mostly strangers," Bill responded when asked about their interactions. While they exercise caution by not adding strangers as friends, the niche community means that "you tend to meet the same people frequently."

For Bill, the allure of VR lies in the ability to traverse borders from the comfort of one's space, emphasizing the joy of "meeting people and being able to 'travel around the world' without having to leave the comfort of my home." However, they're also acutely aware of its pitfalls, acknowledging that "VR can cause you to interact less with 'real-life' friends and family. The opportunity for anonymity can be abused as well." On a more personal note, Bill shared the physical ramifications of prolonged VR usage, admitting, "I noticed that when I use VR for extended periods my eyes tend to feel strained, and I occasionally experience fatigue." Bill provides a window into the evolving landscape of virtual interactions among university students, offering both enthusiasm for VR's potential and a cautious nod to its drawbacks.

Embracing Digital Realms: Insights from an IT Professional

Venturing into the intricacies of VR often unveils diverse perspectives. The fascinating perspective of a person who I will call "Chris" as a pseudonym, a 47-year-old self-employed individual, offers us a glimpse into the blend of profession, passion, and gaming within the sphere of VR. With a career in the IT sector, Chris naturally finds himself on the frontlines of technological advancements. He shared, "I work in the IT sector, so I often have the opportunity to try new technology. That's how I became a gamer and a participant in online social spaces." His foray into VR wasn't just a recreational venture; it inadvertently became a tool for self-improvement. He revealed, "I even learned English this way."

For Chris, VR isn't merely a gaming platform but an evolution in entertainment. He expressed, "I believe that VR allows users to immerse themselves in virtual worlds. Making them feel connected to the gaming, thus improving the entertainment experience by providing a sense of presence and interactivity." Social interactions within VR for Chris have seen a transformation over time. Initially a "silent participant," he gradually found comrades in the virtual world, noting, "Later on, I found a few people that I would play with across different platforms." However, with the increasing influx of younger players, Chris' interaction is often reserved for discussions about game strategy with peers, saying, "Nowadays, there are a lot of kids flooding these spaces... I just play the games and don't participate in the chatting." Enumerating the advantages of VR, Chris praised its ability for "immersion for gamers, and new experiences with playing games different from the past." Yet, he's not blinded by its allure and candidly pointed out the challenges, such as the escalating costs for enhanced experiences and the impracticality of its bulky equipment, commenting, "The more you want to heighten your experience the higher the expenses can be. Also, you have to have enough space. You can't travel with all of the equipment easily."

Shedding light on the physical implications of VR, Chris admitted to occasionally experiencing motion sickness, especially when losing track of time. Chris' narrative provides a mature, balanced view of VR's potential, blending the promise of immersive experiences with the practical challenges they bring.

Exploring the Uses of VR: Insights from a Creative Scholar

When diving into the world of VR, perspectives can be as diverse as they are enlightening. For this person, who I will call "Dylan" as a pseudonym, a 27-year-old Master's student, VR is more than just a gaming tool—it's a canvas for creativity, learning, and exploration. Tracing back to the roots of their VR journey, Dylan reminisced, "I started playing video games when I was in elementary school... And my first VR experience was around five or six years ago." The transition from conventional video games to VR marked a significant chapter in their digital interactions.

While entertainment remains a primary use for Dylan's VR encounters, they recognize its broader potential, noting its potential, "Mainly for entertainment, but also for educational and creative purposes." This aligns with the growing global trend of using VR in diverse sectors, ranging from healthcare to academia. Dylan's social interactions in the virtual world are selective and rooted in real-world connections. They remarked, "I don't interact with anyone online other than a few real-life friends... I find interaction with strangers online cringey and I'm not the biggest fan of it." Their stance reflects a common concern about privacy and the unpredictable nature of online interactions. Dylan was generous in detailing the multiple advantages of VR. Highlighting the blend of entertainment and learning, they said, "One of them is the possibility to 'travel' and see landmarks, and experience virtual tourism from home... people can practice and develop skills in a safe and controlled environment . . . I enjoy that VR offers new avenues for artistic expression and creativity, like 3D drawing and sculptures and other visual non-traditional art forms."

However, acknowledging the challenges, Dylan pinpointed the cost and impracticality for regular travel, observing, "It can be expensive and as a student, who is traveling a lot, it's not possible to take all this technology with me whenever I go." Regarding the health implications, Dylan candidly shared, "I sometimes feel very overwhelmed by the sensory stimulation after using VR for more than 4 or so hours . . . prolonged use of VR headsets, which can be heavy and cause pressure on the face and head, leads to physical discomfort and soreness." Through this conversation with Dylan, we witness the multifaceted allure of VR—a fusion of gaming, artistry, and academic potential, coupled with its inherent challenges and physical demands.

While some see VR as an innovative tool for creative expression, others harness its power for education, social connectivity, or pure escapism. This variance is expected to become even more pronounced as the metaverse continues to grow, evolve, and intertwine with our daily lives. The language of VR, much like the global languages of our planet, is poised to diversify, becoming richer and more nuanced. Our journey into these individual narratives reaffirms a singular truth: The future of VR is as varied and dynamic as its users, and understanding their psychology is key to shaping a metaverse that resonates with humanity's diverse needs and desires.

6

Learning in the Metaverse

Step into the metaverse and you might feel as though you've wandered into a game room rather than a classroom. The vibrant landscapes, interactive characters, and captivating experiences resemble elements of a video game more than traditional education (see Figures 6.1–6.3). This first impression may lead many to view the metaverse as a mere digital playground, suitable only for entertainment, not a serious environment where meaningful learning can happen. However, is this judgment fair, or are we missing the bigger picture?

In the previous chapter, we delved into the engaging nature of the metaverse and how it could transform learning into an enjoyable and anxiety-free adventure. Far from being a frivolous distraction, this immersive and interactive universe can spark curiosity and make education an exciting journey, rather than a tedious task. Yet, questions remain: Is the metaverse just a digital diversion, or does it offer genuine educational value? Can it be a serious space for learning? Is it helpful across various subjects and sustainable in the long term? Are there generational gaps or differences in its effectiveness according to demographic? And how does the rapid pace of technological development affect our understanding? We find ourselves in a time when technology is advancing at such a rate that it often surpasses our ability to fully comprehend its efficacy and potential problems. Previously, research could keep pace with technological development, but nowadays, technology is advancing so swiftly that continuous updates to research are required. Studies are also challenging to generalize, as a lot depends on the particular platform in use. This adds layers of complexity that need careful consideration.

In this chapter, we explore the metaverse's role in education, presenting a series of case studies that highlight its varied applications. It's crucial to note that these studies serve as examples, capturing specific experiences rather than forming a basis for broad generalizations. With the dynamic nature of platforms and the diverse array of users and objectives, it's challenging to draw overarching conclusions. The game-like nature of the metaverse offers engaging experiences, but there's a risk of diminishing the seriousness of academic discussions. The

Figure 6.1 Example of a metaverse classroom. Reproduced with permission of Gather.

introduction of advanced AI models, like ChatGPT, adds another layer of complexity. While they hold potential to enrich learning, their presence also introduces variables that might alter the dynamics and outcomes of educational interactions. The allure of the metaverse lies in its potential for fostering personalized learning. Integrating this rapidly evolving tool into traditional educational practices is a nuanced challenge. My personal explorations, such as employing 2D metaverse platforms to teach the Korean language, will serve as practical examples of its real-world applications. This chapter blends theoretical insights with tangible experiences, aiming to offer a comprehensive understanding of the metaverse in education, with an emphasis on the selected cases being descriptive, not prescriptive.

Educational Reform in the AI Age

Amid the backdrop of rapid technological advancements, the landscape of education has undergone transformative shifts. These shifts not only affect

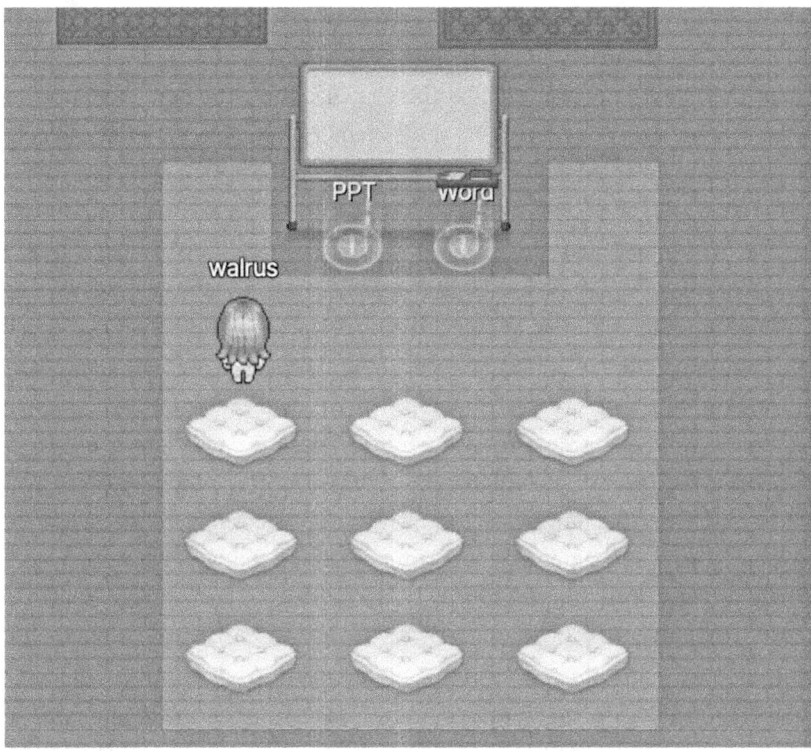

Figure 6.2 Whiteboard and multimedia resources in the metaverse classroom. Reproduced with permission of Gather.

Figure 6.3 Quiz game in the metaverse classroom. Reproduced with permission of Gather.

methods of delivery, with both human and AI educators now at the forefront, but also the spaces where learning takes place. We are transitioning from traditional physical settings to expansive digital and virtual realms. The Covid-19 pandemic played a pivotal role in amplifying these changes, bridging the divide between the physical and virtual, and making the digital transformation of education an immediate reality for many. Textbooks have long held a cherished place in the realm of education, possessing both symbolic and functional value. But as we navigate the shifting terrain of education, particularly in language learning, their relevance is under scrutiny. While hefty language textbooks have their merits, they increasingly face challenges in engaging the tech-savvy Gen Z, as well as Millennials who have tasted the allure of multimedia resources. The pages of traditional textbooks and flashcards, while classic, don't always resonate as they once did, given the myriad of interactive options available today. For many, the pull toward more dynamic, interactive, and tailored learning experiences grows stronger. Teachers recognize this evolving preference but grapple with ensuring they don't lean entirely on technology.

With the advent of generative AI, we may be approaching a harmonious balance between the old and the new. Similarly, the classic classroom structure, while foundational to many of our educational experiences, is undergoing a transformation. The emergence of the metaverse and online education platforms suggests that the four-walled classroom might not remain the norm. These digital spaces offer remarkable accessibility, making education more inclusive and diverse. Preliminary observations hint that within the metaverse, bolstered by generative AI interactions, students might feel a sense of ease and a willingness to explore, ask questions, and participate. For introverted learners, AI-facilitated education could provide a more comfortable and productive environment. While traditional classrooms have their unique benefits, it's evident that the landscape of learning is evolving, blending the best of both worlds.

Metaverse Learning: Navigating the Gray Areas

At its core, the metaverse is like an expansive, immersive video game, but it offers more than just play—it's a new classroom. This digital realm provides innovative ways to learn, shaking up our old views on education. Yet, it's not a simple switch from old to new. In a 2017 study, researchers Dede et al. dove into how VR, AR, and MR can be used in classrooms. They categorize immersive interfaces into VR, multiuser virtual environment (MUVE), and mixed reality. They highlight four immersions that promote learning: actional, symbolic, sensory, and social. They suggest four ways through which students feel deeply involved in learning: doing, connecting to past knowledge, using their senses, and interacting with others. Such tools allow lessons to be more hands-on and relevant, making learning stick better. These methods make education more memorable and tailored.

However, with the fast pace of technological changes, staying updated is a challenge. And while some educators are excited, others remain wary. It's clear: The metaverse isn't a one-size-fits-all answer. There's much to learn and understand. Its potential is vast, but so are its challenges. Rather than rushing in or avoiding it entirely, a thoughtful exploration is crucial. As we move forward, collaboration between tech experts and educators is vital, ensuring we make the most of this evolving landscape while acknowledging its nuances. Beyond offering enriched learning platforms, new technologies introduce methods like real-time simulations, embodied cognition, and immersive narratives that empower students with choice and autonomy. This potent mix of tools offers experiences beyond traditional classrooms, revealing the vast potential within the metaverse. Still, a challenge lies in balancing these advancements while addressing inherent weaknesses. Despite the potential of AI, there is palpable skepticism among educators. The task isn't just understanding the metaverse but also seamlessly integrating it into existing pedagogies. This calls for a collaborative effort between tech experts and educators. They must merge the novel with the conventional, approaching this new learning frontier with hope, but also a healthy dose of pragmatism.

Physical to Virtual Spaces: Tricky Matters

The transition from physical to virtual interactions is not without its challenges, particularly when it comes to overseas students. It brings about a certain level of confusion as we grapple with how to "translate" our physical activities into the virtual realm. In recent years, I received an email from my department requesting confirmation of contact points. My university wanted to ensure that I had sufficient contact with my students, both locally and overseas, and requested that I report any instances of nonattendance for visa and immigration purposes. Yet what about online interactions with overseas students? Do they count as contact points too? Nowadays, even when students are physically located overseas, I often meet them online. In the realm of virtual meetings, does one's physical location truly matter? We are forced to reevaluate our previous notions of physical presence and spatial interactions, especially in the context of engaging with overseas students.

This shift toward virtual education also touches upon some tricky areas. If education can largely occur in virtual spaces, do overseas students need to physically move to learn? Can we all learn from the comfort of our own place by tuning into virtual spaces? These questions challenge the traditional notion of physically attending educational institutions and urge us to reconsider the possibilities of virtual learning. It opens up new avenues for flexible and accessible education, but also poses bigger challenges that perhaps we are not ready to answer.

AI-assisted Learning

AI-assisted self-language learning proves highly effective and sustainable, particularly when compared to traditional self-study methods. The effectiveness of AI-assisted language learning is most notable when learners have access to semi-structured guidelines. In my pilot study, one group of learners engaged in AI-assisted learning independently, while another group received tasks from teachers. Interestingly, the latter group achieved more success. Often, when learners attempt AI-assisted learning without teacher intervention, they may feel somewhat lost. AI-assisted self-study facilitates personalized learning, enabling individuals to customize their language learning experience according to their unique needs and preferences. However, it's important to note that more research needs to be done in this area. This is just the beginning. Importantly, teachers should not view AI as a competitor but as a collaborator. In this era of AI, teachers are more important than ever, but their roles need to evolve. Teachers are increasingly vital in the age of AI, but their roles must adapt to the changing landscape of education.

Teachers' Relationship with Technology

As we navigate the contemporary, technology-rich educational landscape, the importance of a balanced view becomes increasingly clear. Teachers hold a spectrum of opinions on technology, influenced by generational experiences and personal perspectives. While Gen Z educators often seamlessly integrate platforms like Gather and the metaverse into their teaching, their older counterparts may grapple with such advancements. Research, such as a study by Tomczyk et al., provides valuable insights into these varied attitudes by analyzing the views of 873 teachers across Latin America and Europe. The overarching sentiment suggests an openness to technology in education. Still, there's a tapestry of nuanced opinions—some educators embrace smartphones as motivational tools, while others, particularly in regions like Ecuador, exercise caution. These techno-skeptical views are not new. For instance, a 1991 *New York Times* article voiced concerns about televisions potentially replacing educators in classrooms (Hershenson, 1991). Fast forward thirty years, and we're amid similar discussions, this time focusing on advanced devices and AI platforms like ChatGPT.

Eickelmann and Vennemann's (2017) research adds further detail, categorizing educators from ardent "ICT supporters" (26.0 percent) to the more reserved "doubters expressing rejection" (24.0 percent). A myriad of factors, including age, nationality, and subject specialization, shape these opinions. As AI carves a more substantial niche in STEM education, it encapsulates the duality of technological promise and associated reservations. Many educators appreciate AI's transformative potential but harbor concerns about its transparency, long-term implications for teaching roles, and suitability for young learners (Kim & Kim, 2022). This brings us to the heart of the matter: the absolute necessity of a balanced perspective. Historically, the waves of technophobia have been cyclical, appearing with each significant technological innovation in education. However, adopting an extreme view—either unabashedly pro-tech or staunchly against—can blind us to the holistic picture. A balanced approach acknowledges the undeniable benefits of technology while critically addressing its challenges. This equilibrium ensures that we maximize the advantages of technological advancements without overshadowing the invaluable human element of education. The accelerated shift to digital modalities, thanks to the Covid-19 pandemic, further underscores this need for balance. The goal is clear: to create an educational ecosystem where technology enhances and complements human educators, thereby making quality education universally accessible without losing the irreplaceable human touch.

Despite educators' resistance, AI is becoming a prominent fixture, especially in STEM education. Still, some educators, despite seeing AI's benefits, remain wary due to concerns over its explainability, its implications for their roles, and its appropriateness for younger learners. An illustrative study involving an AI tool (AISS) backed by GPT-2 shows this duality (Kim & Kim, 2022). While teachers appreciated its assistance in scientific writing, apprehensions lingered. Current meta-analyses and trends indicate AI's growing role in STEM. Its applications, from enhancing information literacy to creating specialized chatbots, are expanding. As educators grapple with AI's pros and cons, the abrupt pivot to online learning spurred by Covid-19 has added another layer to the discourse. It underscores both the exciting possibilities and the challenges of weaving AI into our educational tapestry.

Gamification Effect

In the evolving landscape of modern education, the term "gamification" has taken centre stage as an influential concept. It signifies the practice of incorporating game design elements into nongaming contexts, thereby enriching learning experiences and fostering engagement and motivation. At its essence, gamification leverages the human tendency toward play, competition, and achievement, transforming conventional learning into a dynamic and interactive process. This transformation is made tangible through platforms like *Roblox*, where the traditional dichotomy between gaming and studying is challenged. The deep-rooted notion that games and education are mutually exclusive is giving way to a more complex understanding. *Roblox* illustrates how the lines between gaming and studying can blur, and in doing so, it brings joy and creativity to the educational process, stimulating engagement and curiosity. Take, for instance, the Math Obby in *Roblox*. Rather than being a simple game, it acts as a catalyst that stimulates mathematical thinking within a captivating and interactive environment. This integration turns rote learning into an appealing challenge, leveraging the natural inclination of students to explore and conquer.

Educators are starting to recognize the value of this approach, seeing games not as a diversion from studying but as a pathway for learning. The collaboration between *Roblox* and various educational institutions reveals a growing acknowledgment of this synergy. However, integrating gamification in education presents challenges. The marriage of entertainment and educational goals requires careful design to prevent the fun factor from overshadowing the learning objectives. Moreover, the long-term effectiveness of platforms like *Roblox* as educational tools is still an area that needs in-depth research and evaluation. The promise is evident, but comprehensive studies are required to gauge the full impact on diverse learning outcomes. The emergence of platforms such as *Roblox* in the educational sphere highlights a broader shift in understanding the relationship between games and learning. It represents a nuanced and multifaceted approach that comes with both opportunities and challenges.

By transforming learning into an interactive and enjoyable experience, gamification is reshaping educational philosophy. The efficacy of this approach within platforms like *Roblox* remains a compelling area of investigation for educators, researchers, and learners. The interplay between game and study is no longer a simplistic contrast but a complex relationship that has the potential to enrich learning on multiple levels. The journey toward this understanding reflects a progressive thinking in education, embracing games not as detractors but as valuable tools for learning and engagement. Given that the information is provided by Roblox, there arises a problem of objectivity in the assessment. While the insights offered are valuable, the unique perspective of the platform may introduce bias. However, with the vast array of other platforms in existence, it's not feasible for researchers to thoroughly evaluate each and every one, adding to the complexity of forming a fully objective and comprehensive view on the matter.

Minecraft Education is revolutionizing the way education is delivered across the globe, enhancing traditional learning with creativity, problem-solving, and digital literacy skills. By developing future-ready abilities like systems thinking, students are enabled to explore real-world issues in immersive and imaginative virtual worlds through game-based learning. *Minecraft*'s in-game coding and tailored curriculum foster computational thinking, preparing young minds for a digital future. Beyond technical skills, the platform also helps students build empathy and learn digital citizenship, contributing to their social-emotional growth. According to Karsenti et al. (2017), educators utilizing *Minecraft Education* report improvements in students' task management, problem-solving abilities, and relationship skills. Its positive impact is being felt worldwide, as supported by integrated curriculum examples and extensive research, cementing *Minecraft Education*'s role as a transformative educational tool.

Finding the Right Balance

The utilization of platforms like *Minecraft* and *Roblox* for educational and creative purposes presents both opportunities and challenges. The message on the role of these platforms must be delivered beyond a black-and-white view. Finding the right balance is crucial. While these platforms offer a new dimension in education, creativity, and social engagement, they are not without drawbacks. Real friendships, physical creativity, and other nondigital activities must not be replaced or overshadowed by virtual interactions. Parents, educators, and policymakers must work collaboratively to ensure that children benefit from these platforms while retaining connections to real-world interactions, friendships, and creativity. It requires continuous monitoring, guidance, and a comprehensive understanding of both the virtual and real worlds children navigate. The success and efficacy of using *Minecraft* and *Roblox* in education greatly depend on how the platforms are integrated and delivered within the curriculum. If thoughtfully implemented, they can become significant assets, engaging students in various competencies and providing fun and interactive learning environments. However, without proper structure, planning, and alignment with educational goals, they can lead to challenges and become problematic. The full educational potential of these games can be harnessed by contributing positively to modern learning experiences. However, it should be accompanied by an awareness of the need to find a balanced approach that does not sacrifice real-world connections and interactions. The debate surrounding these platforms is complex and multifaceted, and the way forward must be paved with care, insight, and a commitment to the holistic development of our youth.

Educational Impact of *Roblox*

Roblox is especially popular among children aged six to sixteen. In 2017, youth under thirteen spent much more time on Roblox than on other entertainment platforms: 2.6 times more than YouTube, 15 times more than Netflix, and 27 times more than Facebook (Tőke, 2019). *Roblox* is more than a platform for playing games; it's a space where children can create their own play spaces. This creative aspect of *Roblox* is indeed attractive for children. It nurtures creativity, encourages problem-solving, and fosters collaboration as they build and explore their virtual worlds. Friends can be found within these digital spaces, expanding social interactions beyond physical boundaries. However, the question arises, Can virtual friendships and creativity replace real-life interactions and activities? Excessive engagement with platforms like *Roblox* may lead to issues. The loss of physical friendships, over-reliance on virtual creativity, and the risk of unhealthy screen time are concerns that cannot be ignored.

Educational Impact of *Minecraft*

Minecraft Education is transforming global education, by fostering creativity, problem-solving, and digital literacy. It develops future-ready skills, allows students to explore real-world issues in immersive worlds, and enhances computational thinking. Students also build empathy and learn digital citizenship, thus developing social-emotional skills. Teachers witness improved task management, problem-solving, and relationship skills, and their impact can be seen worldwide. The pedagogical use of *Minecraft* was evident in a study involving 118 students in Québec (Karsenti et al., 2017). The program featured thirty tasks divided into ten progressive levels, aimed at developing learning skills like motivation, computer programming, basic informatics skills, and peer collaboration. A total of twenty-five benefits were identified, including stimulating and sustaining student engagement. Countries like Sweden have integrated *Minecraft* into the curricula, recognizing its significant pedagogical benefits. The same study suggested a pronounced increase in student motivation. For instance, a school principal received an email from a student's father, indicating that even when school was out, his daughter was eager to return for *Minecraft*. Notably, despite these *Minecraft* sessions being voluntary and held after school, the moderator reported high attendance, pointing to students' pronounced enthusiasm and interest. The demand was so high that a principal had to limit participation. Furthermore, surveys from the study revealed that 77.1 percent of the students found playing *Minecraft* at school "extremely" enjoyable, emphasizing their joy in constructing cities and various objects and valuing the game's creative and educational components. However, while these findings are undeniably positive, it's crucial not to overgeneralize. The interpretation of such data is vital, as the effectiveness of tools like *Minecraft* can vary based on context. Deciding on the degree and timing of its integration in educational settings is a nuanced process and needs tailored considerations for each application.

Challenges and Recommendations: A Look at *Minecraft* in Education

Minecraft's incorporation into educational settings has brought about certain benefits. On the other hand, challenges have also been observed. There have been instances of technical issues, and there's an inherent need for computers that align with gameplay specifications. This has highlighted the pivotal role of a knowledgeable moderator or teacher, someone familiar with the intricacies of gameplay and computer functionalities. A financial component also exists in this equation. Access to the educational version of *Minecraft* requires a Microsoft Office 365 account, priced at 5 USD (6.73 CAD) for single-user subscriptions. This could present affordability concerns for students with limited financial means.

Drawing from Karsenti et al.'s (2017) findings, several suggestions arise for a smoother integration of *Minecraft* in educational contexts. Schools could consider facilitating structured yet flexible *Minecraft* sessions, ideally under the guidance of a qualified moderator. Tailored tasks and progressive levels might be effective in keeping student engagement levels high. By providing students with some freedom in gameplay while simultaneously encouraging teamwork, a balanced learning environment could be fostered. It would be advantageous for students to share and discuss their experiences, promoting clarity and reflective thinking. While recognizing students' achievements within the game can be beneficial, it's essential to remember that *Minecraft*, at its core, is a recreational game. Regarding the research's objectivity, caution is advised when interpreting results, especially if the source might have interests in the technology discussed. Even if the research methods are grounded in neutrality, there remains a possibility that not all facets, especially potential disadvantages, are adequately represented. To gain a well-rounded understanding, assessments from independent research entities could be beneficial. Still, the rapidly changing nature of technology and the emergence of new platforms make this a complex endeavor.

The findings detailed are primarily based on a report that highlights the positive impact of *Minecraft* on educational outcomes. While most studies referenced present similar results, it's important to approach these findings with caution. The success and efficacy of using *Minecraft* in education greatly depend on how the platform is integrated and delivered within the curriculum. If used thoughtfully and strategically, *Minecraft* can become a significant asset. However, if implemented without proper structure, planning, and alignment with educational goals, it can lead to challenges and become problematic. Educators, parents, and institutions must therefore carefully examine the best practices and approaches for using this platform, ensuring that it aligns with pedagogical principles and the unique needs of their students. By doing so, the full educational potential of this game can be harnessed, contributing positively to modern learning experiences.

Metaverse versus Conventional Platforms

Virtual education is in the throes of a paradigm shift. While platforms like Zoom have been instrumental in the early stages of online learning, emerging preferences among learners suggest a burgeoning inclination for metaverse classrooms. The crux of the issue with Zoom and similar platforms lies in their limitations in promoting genuine interactivity and the challenges they pose for small-group discussions. On such platforms, even PowerPoint presentations can feel isolating, akin to casting words into a void without sensing the pulse of the audience. Contrarily, the metaverse, especially in contexts like teaching Korean, emerges superior in several areas: time efficiency, reduced psychological strain, heightened engagement, enriched pair work opportunities, and fostering a more organic teacher–student rapport. The metaverse's advantages include tools like virtual whiteboards that allow instantaneous monitoring of students' tasks, facilitating prompt feedback. If real-time observation proves elusive, the work remains archived for later perusal. This dynamism extends to breakout areas, designed to hone focus by limiting interactions, thus reducing potential distractions. Such spaces empower students while also allowing teachers to monitor subtly.

In the metaverse, reticent students often find their voice. Digital tools like chat boxes provide alternatives to traditional voice and video inputs, alleviating the pressures of being "on display." The presence of avatars acts as a bridge, enhancing personalization and fostering a collective identity, even when traditional video conferencing feels distant. Flexibility is another commendable trait of the metaverse. Catering to diverse backgrounds and conditions, it democratizes the learning space. Students can immerse without fretting over externalities. For educators, the 2D design is a boon, eschewing the disruptions associated with VR tech. This design, while tech-forward, also promotes an egalitarian teaching ethos. The absence of commutes in the metaverse is a resounding advantage from the student's lens, freeing up invaluable time for both academic and personal endeavors.

As the metaverse surges forward, platforms like Zoom show their age. In my interactions on Zoom, the lack of visual feedback is a conspicuous limitation. Video meetings, while revolutionary, lack the tangible energy of in-person interactions, resulting in a palpable disconnect. For someone who often hosts conferences, this disjunction is even more glaring. While physical spaces offer spontaneity and organic audience engagement, their virtual counterparts fall short. Given these observations, it's clear that a mere fusion of online and offline modes—the hybrid model—isn't the panacea we once hoped for. A more immersive, integrated platform seems imperative. Reflecting on the 2020 to 2021 remote instruction period, the debate around mandatory camera usage on platforms like Zoom exemplifies existing challenges. While continuous camera usage sparked "Zoom fatigue" and privacy concerns, a camera-off policy threatened to dilute the learning experience. The balance may well lie in innovative solutions like avatars or personalized icons, which encapsulate the ethos of presence without the pitfalls of real-time video. This might just be the equilibrium we seek: a blend of engagement, comfort, and privacy that charts the future of online learning.

Metaverse and Robots: Aiding Social Skills Development in Autistic Children

The recent MIT Technology Review sheds light on the potential benefits of metaverse environments and robotics for children with special needs, specifically those diagnosed with autism spectrum disorder (ASD) (Hao, 2020). Autism is a developmental condition that impacts approximately 1 in every 160 children worldwide, presenting challenges in social, emotional, and communication areas. The therapy required for autism can be expensive and time-consuming, often requiring around 20 hours of therapy per week, making it imperative to explore more accessible solutions. Technology, in the form of socially assistive robots and immersive metaverse environments, has emerged as a promising tool to provide these solutions.

In metaverse spaces, children with special needs can interact in a controlled yet dynamic environment that encourages exploration, social interaction, and learning at their own pace. Similarly, in-home socially assistive robots have the potential to offer personalized care to autistic children, addressing their unique needs and symptoms. A recent study highlighted an AI model capable of predicting a child's engagement during therapeutic activities involving these robots with 90 percent accuracy (Jain et al., 2020). The study took place over a month-long period in which the robots lived in the homes of autistic children. The robots engaged the children in space-themed math games. However, the main focus wasn't the games themselves, but fostering the development of fundamental social skills through these interactions. Many children came to view the robot as a friend, leading to noticeable improvements in their empathy toward peers and family members. Despite the encouraging results, the in-home study encountered challenges, including inconsistent and noisy data due to accidental damages or interference from siblings. Privacy concerns also arose, prompting researchers to explore ways to minimize the data needed to train the robot's machine-learning algorithms effectively. The hope is that these technological advances in the form of metaverse spaces and socially assistive robots will become accessible, affordable, and personalized therapeutic aids for children with special needs, especially those with autism. These tools could lead to more comprehensive care and improved developmental outcomes.

Hybrid Education: A Blend of Traditional and Digital Learning

Education is continuously adapting. Education is now evolving toward the hybrid model, combining the strengths of online and offline teaching. A prime example is Minerva University, which, since its inception in 2012, has blended the virtual seminars of its "Active Learning Forum" with the tangible experiences of its "Residential Semesters". Here, students are enriched with both online academic insights and real-world exposures in diverse global cities. Similarly, we can see the duality of hybrid conferences, in which participants from across the globe are able to participate, breaking down geographical barriers and allowing those who may have struggled to attend due to other constraints—such as financial or familial—to join in. That said, while they have their advantages, they are far from a perfect model. For instance, access to these conferences hinges significantly on digital equipment, highlighting and exacerbating the divide between users who have better equipment and those who do not. My own experience of organizing a conference on languages in Asia from an AI perspective has led me to see the positives of an entirely virtual conference, rather than a hybrid one. In particular, a conference where everyone is online levels the playing field; no single group will be left out or waiting awkwardly in the delay between those physically at the venue and those online—everyone is equal. In this sense, diving fully into the metaverse also has its pros in comparison to a blended approach. In addition, from my own experience teaching Korean, I've found the metaverse to be an invaluable tool, particularly when tackling pragmatics. Traditional methods or even app-based lessons aren't always sufficient to convey the nuances of pragmatics. However, in the expansive universe of the metaverse, where students interact as avatars, these challenges can be addressed in a novel manner.

Yet, the metaverse also has its limitations. Spontaneous interactions and the kind of empathy-building that often comes naturally in traditional settings can sometimes feel constrained in a purely digital environment. Furthermore, for tasks that require teacher intervention, such as real-time writing instruction, the tactile, immediate feedback in an in-person setting can prove more effective.

Considering these findings, it's evident that combining the strengths of the metaverse with the irreplaceable elements of face-to-face interactions offers the best results. As technology continues to evolve and embed itself in the education sector, it is likely that the hybrid approach will become a predominant trend, shaping the future of teaching and learning across various disciplines. However, determining the right balance in this hybrid approach largely depends on each educational environment, presenting a challenge that future educators must address and refine.

Metaverse and Co-learning

The metaverse's unique structure adds another layer of possibility to the learning environment by often obscuring traditional markers like age, background, and other personal characteristics that might be known in a physical classroom. In the metaverse, participants may engage with each other without immediate knowledge of these factors, allowing interactions to be more focused on ideas, skills, and contributions rather than preconceived notions or biases related to age, appearance, or background. This anonymity, whether partial or complete, can foster a more open and equitable discussion. For example, a younger student might feel intimidated expressing their ideas in the presence of older, more experienced learners in an in-person classroom setting. In the metaverse, however, these age differences may become irrelevant, allowing the younger student to participate more fully and confidently.

Anonymity in online spaces can also eradicate the judgments that may arise from appearance. In the metaverse, participants may be able to approach each other's ideas without bias, leading to a more genuine collaboration and the inclusion of a broader range of perspectives. This aspect of the metaverse can help create a more inclusive and engaging learning environment where collaboration is encouraged, and ideas are judged on their merit rather than the characteristics of the person presenting them. Teaching in the metaverse can shift the educational paradigm toward a more egalitarian and participatory model, fostering an atmosphere of co-learning that is rich in diversity and free from traditional constraints.

2D Metaverse in Education

Metaverse communication has garnered attention for its ability to transcend the barriers of traditional virtual platforms. While the 3D metaverse, often reliant on VR headsets, has limitations—notably, its short span of comfortable usage—the 2D metaverse is a more versatile alternative. A more detailed list of benefits and challenges of the 2D metaverse can be found below.

The Advantages of 2D Metaverse Communication

- Avatar-Based Interaction: Platforms like Gather Town leverage on avatars to offer a mix of virtual and face-to-face interactions. Avatars eliminate biases based on physical appearance and enable unhindered self-expression. They evoke feelings of presence and vibrancy, making one feel closely connected with peers and professors.
- Reduced Psychological Distance: Unlike typical video conferencing platforms, such as Zoom or WebEx, 2D metaverse platforms don't pressurize users to be constantly on camera. The use of avatars alleviates feelings of self-consciousness, enabling more spontaneous conversations.
- User Accessibility: 2D platforms run seamlessly on common browsers like Chrome, making them widely accessible without steep learning curves.

Challenges of 2D Metaverse Communication

- Technological Adaptation: Not everyone is familiar with the nuances of 2D platforms. The learning curve required to gain competency can create gaps in participation and engagement.
- Balancing Real and Virtual: As an intermediary between real-life and virtual reality experiences, metaverse communication may not fully encapsulate the profound advantages of either.
- Graphical Interpretation: Some users may find the gamified visuals of the 2D metaverse inappropriate for academic deliberations, craving a more serious or scholarly backdrop.
- Mobile Engagement: Interaction on mobile devices can be less intuitive, which might affect participation.
- Depth of Engagement: The intimacy and granularity of offline interactions might be missed. Voluntary participation could be lower than in compulsory offline contexts. However, emerging research underscores the advantages of immersive digital platforms. For instance, studies indicate that platforms like Second Life enable richer teacher–student interactions as compared to traditional online learning avenues. This fosters a sense of unity and narrows perceived distances.

The ongoing debate regarding online versus offline communication doesn't conclude that either one is superior. Much hinges on the specific context, indicating a future where proficiency in both modes becomes essential. It's not so much about one form of communication being vastly superior or achieving better outcomes. Rather, in-person and virtual communication are likely to coexist, evolving side by side. They don't seem to be on paths where one overtakes the other, but instead, each influences and enriches the other, expanding the diverse spectrum of human communication.

Possible Distractions

The metaverse presents a compelling proposition for the domain of education. As we discussed earlier, it provides avenues for immersive learning environments, gamified experiences, and a platform where education could undergo significant transformation. The appeal of the metaverse stems from its ability to offer novel ways to engage learners in environments that were previously limited to the imagination. Yet, alongside its potential, there are challenges that educators, students, and parents should be aware of. Distractions, as highlighted by Chu et al. (2021) in the context of mobile phones, may become even more pronounced in the metaverse. The vastness of this virtual realm can sometimes blur the boundaries between educational content and other activities that might not directly contribute to learning. The metaverse, in many ways, is uncharted territory for education. Its capacity to redefine learning is evident, but it also introduces variables that can hinder the learning process. For educators and parents, the task becomes one of discerning meaningful experiences from potential distractions within this space. However, turning our backs to the metaverse is not the solution. Its presence in the contemporary digital landscape is undeniable and growing. The approach should be one of understanding and engagement, where parents and educators familiarize themselves with its offerings, acknowledging both its potential and its challenges. Wrapping up, the metaverse brings to the table both exciting opportunities and inherent challenges for education. It's essential to recognize its dual nature: offering innovative learning possibilities, but also demanding careful navigation to ensure productive educational outcomes. Engaging with the metaverse means actively exploring its potential while being mindful of its challenges to create a balanced educational experience.

AR: Boosting Motivation and Technological Challenges

Bower et al. (2014) investigated the use of AR in both society and educational contexts. They highlighted the advantages of AR for teaching, particularly in terms of increasing student motivation and fostering higher-order thinking skills. The authors discussed the technologies required for AR implementation, such as GPS and image recognition software. While AR has shown promise in various fields, including science and humanities education, it still faces barriers to widespread adoption due to technical challenges. Despite the potential benefits, the complexity of technical support remains a significant obstacle in leveraging AR effectively for educational purposes. Boyles (2017) looks at the advantages, drawbacks, and applications of AR and VR in education, specifically examining how these technologies are applied to teaching at an American military academy. He notes that VR and AR are used in many different facets of education, from medical education to engineering education and distance learning. In particular, VR has been employed in foreign language learning to enable learners to interact with native speakers. Several advantages of VR are suggested, including enabling teachers to not just transmit knowledge but actually facilitate student-led independent learning that empowers students, and how VR can assist students in visualizing abstract concepts and help them to perceive objects on various scales (of size). That being said, VR also has its drawbacks, as it can be expensive and require a lot of technical paraphernalia. Teachers and students may need a lot more time to learn how to use VR, which necessitates training and potential technological malfunctions. Applying AR and VR specifically to the military education setting, Boyles notes that these technologies can help cadets to learn military history through reenactments and engage in combat simulations. This allows them to experience dangerous situations remotely without risks.

AR in Education: Benefits, Challenges, and Evolving Trends

Garzón et al. (2019) analysed sixty-one studies on AR in education from 2012 to 2018. The research found AR mainly benefits children, teenagers, and undergraduates in subjects like natural sciences and mathematics. It enhances learning, motivation, memory retention, and learning autonomy, although special needs users were considered in only 2.5 percent of the studies. Drawbacks included complexity for users and resistance from educators. It's worth noting that, as technology advances, these findings may change. More research is needed, and trends in A's educational impact are likely to evolve in the coming years depending on technological developments.

Akçayır and Akçayır (2017) conducted a comprehensive literature review on the use of AR in educational settings. Their analysis of sixty-eight articles examined the benefits, challenges, types of learners (such as K–12 students and adult students), time period of study initiation, and employed AR technologies. The research revealed a gradual increase in AR adoption since 2007, with a significant surge of interest starting in 2011 due to the emergence of mobile devices. The primary focus of the examined studies was on K–12 students and university students, with mobile devices identified as the most popular AR technology due to their portability and interactivity. The authors emphasized the importance of technologies that most "clearly supports meaningful learning" (p. 5). The advantages of AR in education were categorized into learner outcomes, pedagogical contributions, interactions, and others, including improved student achievements, increased enjoyment in classroom instruction, enhanced student-to-student interaction, and visualization of abstract concepts. However, AR presented challenges, including the difficulty of using AR technology and the increased cognitive load associated with complex tasks in an AR environment.

Bacca et al. (2014) analyzed thirty-two articles on the use of AR in education. They found that AR was commonly used in science education, followed by the humanities and arts, including language learning. AR was applied in higher education, as well as primary and secondary schools. The main functions of AR were explaining topics and providing supplemental information. The benefits of AR included increased motivation and learning gains, while the difficulty in accessing augmented data was a significant drawback. AR improved learner performance and motivation. There was a lack of studies tailored to special needs students. The authors emphasized the need for accessible and personalized AR options to cater to diverse user needs.

In their literature review spanning 2010 to 2016, Dalim et al. (2017) identified six key factors shaping the acceptance of AR in education. These factors include alignment with curriculum and pedagogical needs, technical stability, support for constructionist pedagogy, consideration of parental involvement, adaptation to diverse student backgrounds, and suitability of the AR platform. A recurring theme among user expectations is the importance of AR in supporting visual learning, especially for grasping spatial and conceptual knowledge beyond traditional text-based methods. Diegmann et al. (2015) conduct a review of the past and current literature on AR, analyzing the benefits of AR in education and how the benefits differ by type of AR technology. The authors note five main directions of AR applications: discovery-based learning (using provided information to discover and think about a certain place or object), objects modeling (visualization of how an object looks like in certain setting), AR books (books augmented by AR technology), skills training (training in specific skills through AR), and AR gaming (AR games in education). Through an analysis of twenty-five articles, they derived a total of fourteen different benefits of AR in education, which were grouped into six main groups including teaching concepts, learning type, and content understanding. Individual benefits in these main groups include increased motivation, improved collaborative learning, increased information accessibility, improved learning curve, improved development of spatial abilities, and reduced costs of an AR-based education as compared to traditional education in the long term. In particular, an improved learning curve and increased motivation were mentioned the most among all the benefits. These articles were then mapped onto the five main directions of AR applications based on their mentioned benefits. Discovery-based learning was the most prolific, as nine benefits and eight articles were mapped onto it.

Koutromanos et al. (2015) reviewed seven papers on AR game usage in formal and informal learning from 2000 to 2014. They found over ten AR applications on iOS and Android, focusing primarily on science education at the secondary and primary levels. This significant variation in applications during the earlier period is surprising. The studies employed mixed qualitative and quantitative approaches using smartphones and tablets. Prominent AR systems included location-based and image-based approaches, exploring constructivism, game-based learning, and collaborative learning. The benefits of AR gaming encompassed scientific argumentation, problem-solving, concept understanding, and student engagement. However, skepticism persists regarding the integration and extent of gaming in education, highlighting the importance of thoughtful integration and diverse implementation approaches.

Lee (2012) explores the applications of AR in both educational and business contexts. In education, AR's use ranges from K–12 to higher education, with applications in subjects like astronomy, biology, mathematics, and physics. Lee predicts future benefits of AR, such as enhanced interactivity in learning environments, contextual quality improvements, and increased educational efficiency, as well as consider AR's potential in safety education. Maas and Hughes (2020) review the applications of VR, AR, and mixed reality in K–12 educational contexts, drawing from twenty-nine studies conducted between 2006 and 2017. The authors define VR, AR, and mixed reality, but note that these definitions can be vague. They find previous literature reviews focused on learning outcomes, instructional design, and the pros and cons of VR and AR but overlooked mixed reality and K–12-specific outcomes. Their review identifies seven key themes: attitude, motivation, engagement, performance outcomes, critical thinking, collaboration, and communication. Examples include positive student attitudes toward VR and AR, increased attention and participation with AR, and enhanced critical thinking and collaboration. The authors highlight a preponderance of AR studies, often involving software and application design, and conclude with a call for more varied research methods, including the exploration of VR through smartphones in K–12 settings, to better understand subject material that yields greater learning gains compared to traditional instruction.

Young Children's Digital Exposure

In the modern era, children's exposure to e-media is virtually unavoidable. Niiranen et al. (2021) conducted a study in Finland between 2011 and 2017. They examined the effects of e-media on 699 children aged five. It found that while 95 percent exceeded the daily recommended usage, leading to a risk of psychosocial symptoms, the impact was not uniformly negative across all platforms. High-dose electronic game use was associated with fewer risks than program viewing. This nuanced picture indicates that the black-and-white view of e-media as solely good or bad is unhealthy and oversimplified. The key lies in how children are exposed to e-media, with platform-by-platform consideration being essential. Health professionals play a crucial role in guiding parents, not to eliminate e-media but to manage it wisely to foster healthy development, recognizing that it can be both beneficial and detrimental.

Oranç and Küntay (2019) conducted a thorough review of previous research, examining how children establish connections between the virtual and real worlds during their learning experiences. They also provided valuable recommendations for employing AR to support early childhood education. The authors outlined four key principles essential for creating effective educational mobile applications that utilize AR. First, they suggested allowing children to interact with both physical and digital tools. Second, they emphasized the importance of engaging and motivating children through AR for better learning outcomes. Third, they advocated for enabling children to bridge the gap between the virtual environment and their physical surroundings, facilitating more meaningful learning experiences. Lastly, they proposed creating a collaborative environment that encourages discussion and active participation. Although various studies have explored AR usage in early childhood education, particularly in areas such as alphabet learning and spatial skills, the authors observed a lack of specific focus on pre-schoolers. They highlighted the significance of understanding how children navigate mixed realities to ensure the success of AR applications. Regarding AR's impact on children's learning, the authors emphasized two critical aspects. First, they stressed that AR should guide children's attention to learning materials and encourage self-reflection while keeping them connected to the real world. Second, they emphasized the importance of seamlessly blending physical and virtual representations to minimize the disparities between the learning context and real-life experiences.

Navigating a Digital Dilemma

Today, my daughter's school provided a digital safeguarding recommendation to parents, highlighting the crucial need to prioritize online safety. The recommendation advises blocking apps like WhatsApp, Hulu, Reddit, Vimeo, Zoom, Xbox Live, TikTok, Tinder, Facebook, Snapchat, Fortnite, Instagram, and Twitter for my eleven-year-old child. The recommendation also suggests that apps, such as TikTok, Facebook, Snapchat, Instagram, Fortnite, League of Legends, and Roblox, can be accessible to children from the ages of thirteen years and up with controlled access. It's important to recognize that the appropriateness and age restrictions of these apps may vary, intensifying the dilemma. However, this recommendation underscores the significance of implementing protective measures to ensure a safe online experience for our children. Striving for a delicate balance between their desire for digital engagement and our parental concerns, we must actively monitor and set limits on app usage based on age and platform.

A Twenty-First-Century Skill

Papanastasiou et al. (2019) conducted a literature review of current studies on VR/AR applications in K–12, higher, and tertiary education to examine how these technologies cultivate twenty-first-century skills and overall generalized learning. The authors begin briefly by looking at what VR is, as well as the importance of twenty-first-century competencies, such as creativity and social skills. They then move on to outline the effects of VR/AR on a wide range of twenty-first-century skills such as retention and memory, collaboration, visuospatial skills, and emotional skills, as detailed in preceding studies. For example, pertaining to collaboration and social skills, virtual environments such as Second Life can promote open, collaborative learning by allowing students and instructors to have a shared experience in the sense that they can work together in classroom instruction in a collaborative style by joining social networks. However, the author also notes the limitations of AR/VR as detailed in previous studies, such as technical drawbacks like bandwidth limitations, which can affect the experience of a 3D environment, and user challenges such as physical ailments (such as dizziness) experienced while using AR/VR equipment and training teachers to teach using AR/VR.

Yuen et al.'s (2011) literature review, conducted over a decade ago, explored the then-recent developments in AR, its societal impacts, and its educational applications. At that time, AR was positioned on a spectrum between real and entirely simulated environments, and the research was primarily focused on device development and application creation. The authors categorized AR's use in various fields such as advertising, architecture, and medicine, and identified five principal educational applications: AR books, gaming, discovery-based learning, object modeling, and skills training. Tools for creating AR content were also mentioned, and the authors concluded with predictions about AR's future in education. However, it is worth noting that in the rapidly evolving field of AR, a time span of five to ten years can bring substantial changes. Research and technology that were cutting-edge a decade ago may now be outdated. As the AR landscape continues to evolve, with constant advancements and innovations, some of the specific tools, applications, or predictions made in Yuen et al.'s review might no longer be as relevant in today's context. The pace of change in this area underscores the need for continuous study and adaptation to keep abreast of the latest developments and possibilities in AR.

AR and VR in Language Learning

Bonner and Reinders (2018) take the first step in this exploration by investigating AR and VR in ESL teaching. They provide insights into the unique affordances such as immersion, portability, and personalization of AR/VR environments. A significant part of their work is about building together a learning environment and connecting the physical body with the virtual. This creates innovative activities like campus tours and treasure hunts. However, they also express concerns about privacy and security, laying the groundwork for a discussion on potential challenges. Cai et al. (2022) delve into the effectiveness of AR through a meta-analysis of twenty-one studies conducted between 2008 and 2020. Their statistical analysis supports the idea that AR's interactivity and multimodal features have a positive influence on language gains and motivation. This affirms Bonner and Reinders' emphasis on immersion and extends it to highlight how short-term exposure to AR can be particularly motivating for elementary students. The study also opens avenues for integration with game-based and problem-based learning, connecting to broader pedagogical approaches.

AR in Early Language Learning

Further expanding on the design aspects, Fan et al. (2020) systematically reviewed fifty-three papers to analyze the design strategies and instructional effectiveness of AR in early language learning. Their review categorizes AR activities and uncovers effective combinations of design and instructional strategies, connecting back to the concept of immersion and personalization. Adding a practical dimension, Hadid et al. (2019) investigate a specific AR project called Reading Buddy, designed to assist English learners. This study not only resonates with the benefits highlighted in the previous works but also gives concrete examples of how AR can lower cognitive processing demands and immerse learners in real-world situations. By introducing Reading Buddy, they offered a tangible solution that leverages AR's benefits and opens up the conversation about potential applications in various learning stages.

AR in Language Teaching

The integration of technology into the educational sphere has been a double-edged sword, and augmented reality in language teaching serves as a prime example of this dynamic. AR, with its potential to revolutionize the educational landscape, brings with it both clear advantages and discernible challenges.

The rise of mobile learning, facilitated by AR, has democratized access to educational content. With just a portable device, students can embark on immersive linguistic journeys—a feature highlighted in various studies including those by Punar Özçelik et al. (2022). Moreover, AR has a unique way of making students active participants in their learning, amplifying their role, and enriching their experience. This active participation, combined with the technology's inherently engaging nature, can bolster motivation and reshape students' attitudes toward learning (Hein et al., 2021; Iqbal et al., 2022; Punar Özçelik et al., 2022). From a pedagogical perspective, the experiential and contextual backdrop AR offers can be a game-changer. Language isn't just about words; it's about context, and AR, through its immersive environments, provides just that (Punar Özçelik et al., 2022).

While AR in education has shown promise, it's essential to note that it isn't without its drawbacks. A central concern is the inherent technological requirement of AR. Students who don't have access to the necessary devices are inadvertently left out, a point emphasized by Punar Özçelik et al. (2022). Moreover, some AR applications are reliant on internet connectivity, which might be inconsistent or even unavailable in certain areas, as highlighted by Nisiforou et al. (2021).

One particular aspect of AR that warrants further exploration is the so-called "wow" or novelty effect. This refers to the initial surge of enthusiasm and improved performance when students are introduced to new technological elements. It's not necessarily because they're learning more efficiently, but rather because they're engrossed by the new, exciting technology. Imagine the enthusiasm when someone tries out a new gadget for the first time—that's the novelty effect in action. While it can lead to a temporary spike in attention and engagement, it doesn't guarantee sustained interest. Over time, what was once novel can become mundane. For instance, the enhanced engagement seen with automated workbooks doesn't always imply that they are superior in fostering long-term learning compared to traditional ones (Hein et al., 2021). Moreover, from a psychological standpoint, anything new or perceived as "threatening" can trigger an initial stress response, which usually fades over subsequent exposures. This isn't just limited to AR but is a general human response to novelty. The challenge for educators and researchers is to differentiate between genuine learning enhancements and temporary boosts due to the novelty effect. Notably, this effect can also distort the outcomes of research studies because participants might behave differently in novel situations, as they might be swayed by the technology's appeal rather than its instructional value (Hein et al., 2021). Furthermore, the novelty effect isn't the only concern with AR. It's critical that both the users, usually the students, and the creators, like educators or app developers, are adept at leveraging the technology. Without the necessary technical expertise, the full potential of AR as a learning tool might remain unrealized, as pointed out by both Punar Özçelik et al. (2022) and Hein et al. (2021).

On one hand, AR promises enriched, context-driven, and engaging language learning experiences. On the other, it brings to the forefront issues related to accessibility, distraction, and a potential skills gap. What's clear is that for AR to truly shine in language teaching, a balanced approach that acknowledges both its potential and its pitfalls is essential.

Virtual Immersion: Making the Year-Abroad Experience Sustainable and Affordable

Immersive language learning has always held a special significance in the domain of foreign language acquisition. At the heart of this concept lies the "year-abroad" approach, where students immerse themselves in a foreign country to learn a language in its native context. This method not only offers unparalleled situational learning but also grants learners an in-depth cultural experience, fostering a deeper connection to the language. However, such traditional year-abroad programs come with substantial challenges. Foremost among them is the cost. With the average semester abroad costing $16,368 in 2023, many potential language learners may find themselves priced out of this invaluable experience (McNair, 2023). Beyond the financial concerns, there are unpredictable barriers like global pandemics, reminiscent of the Covid-19 crisis, that can halt international travels indefinitely. Additionally, frequent international travel carries environmental concerns, adding to the carbon footprint in an age increasingly conscious of sustainability.

This is where the transformative potential of virtual immersion comes into play. With the advent and proliferation of metaverse platforms, learners can now experience a virtual "year abroad" that simulates the cultural and linguistic immersion of the traditional experience at a fraction of the cost. Not only does this make language learning more accessible and affordable, but it also offers a more sustainable alternative, circumventing the environmental concerns associated with frequent air travel. In conclusion, as technology continues to evolve and redefine boundaries, the metaverse and virtual immersion could very well be the future of immersive language learning, making the cherished year-abroad experience both sustainable and affordable for all.

Foreign Language Learning with Immersive Technologies

Language learning has traditionally thrived on immersion, plunging the student into environments where the language pulses with vitality and intertwines with everyday experiences. This immersion often implied travel, residing amid native speakers, and soaking in the rich tapestry of a foreign culture. However, the dawn of technology is reshaping this paradigm. The metaverse holds the potential to revolutionize the concept of "year-abroad" language studies in the future, making immersive language learning experiences more accessible and affordable for many. The emphasis on gamification opens doors to new forms of interaction and collaboration, encouraging learners to explore language in real-world contexts, even within a virtual space. The potential benefits extend beyond mere motivation, offering possibilities for long-term retention, cultural understanding, and practical application of language skills. The rising trend of gamification and immersive technologies in language education underscores the need for ongoing research, development, and thoughtful integration into curricula. The adaptability of these tools to different languages, age groups, and learning objectives makes them versatile and appealing, yet it also calls for careful consideration of the platform and particular situations to generalize the pros and cons. This trend will continue to grow, offering opportunities and challenges for educators, developers, and learners alike.

Hein et al. (2021) bring a broader perspective by analyzing fifty-four articles on immersive technology in foreign language learning from 2001 to 2020. Their work synthesizes the areas in foreign language learning where immersive technologies are used and how they impact learning outcomes. This study adds a layer of understanding by revealing under-researched aspects like affective factors and teacher effects. It also complements Fan et al.'s (2020) work by identifying gaps and areas for future research. These studies together form a comprehensive picture of AR and VR's current landscape in language learning. They reveal the significant advantages, such as immersion, personalization, and motivation, with the latter being particularly impactful for younger learners. However, the research also sheds light on the fact that these motivational aspects might differ among various age groups and could potentially be less effective for older individuals. The collective insights also raise important considerations about educational equity. While digital education, including AR and VR, presents an opportunity to democratize learning, the uneven access to and familiarity with these technologies may lead to inequality and injustice. Challenges such as privacy concerns, technical difficulties, and the specific needs of teachers and students add complexity to the integration of AR and VR in language learning. The interconnection between these studies emphasizes the multifaceted nature of this field and the need for a holistic approach, considering not only the technological advancements but also the human, ethical, and societal dimensions.

Immersive technologies like AR and VR have become central to modern language education. I will present some meta-studies here that delve into the multifaceted role of immersive technologies like AR and VR in the realm of language education according to various literature reviews conducted between 2015 and 2021. Huang et al.'s (2021) examination of eighty-eight articles reveals how AR and VR are applied in language learning, highlighting immersion as the key method. College students were found to be the main users, with the technologies boosting learning outcomes, motivation, and reducing anxiety. The study also indicates the need for more teacher training and the exploration of learner-centered factors. Karacan and Akoğlu's (2021) review emphasizes the different types of AR applications and their affordances in language education. They detail applications such as image-based AR, markerless AR, and creation-based AR. They outline the potential benefits of AR in enhancing authentic learning, motivation, information retention, and language skills, while evaluating various factors like teacher and student perspectives, learning theories, infrastructure, and sustainability. They warn, though, that since AR was not specially developed for education, integration may require more effort from teachers. Majid and Salam's (2021) work uncovers trends in AR usage in language learning over a decade, finding an increasing demand for AR-related research in language learning. Their study shows that the majority of AR research in education is related to science, with language being a smaller focus. They note various common techniques and language skills, such as marker-based AR, mixed research design, and the focus on vocabulary, orthography, and pronunciation.

Parmaxi and Demetriou's (2020) comprehensive analysis of fifty-four articles explores various devices, languages, and contexts used for AR applications, and how AR learning corresponds to twenty-first-century skills. Their findings highlight the usage of mobile-based AR, languages like English and Chinese, varying research participant numbers, and the duration of AR activities. They provide recommendations for future AR research, emphasizing the development of standardized methodologies and integrating meaningful, real-life situations. Similarly, Parmaxi's (2023) review of twenty-six studies on VR in language learning from 2015 to 2018 reveals the use of platforms like Second Life and the lack of studies on fully immersive VR systems. The review identifies

English as the most investigated target language and emphasizes speaking as the most studied skill. The author notes the challenges in implementing VR due to technicalities and pedagogical grounding.

Face-to-Face Versus Virtual Communication in Learning Environments

As our world becomes increasingly digitized, a full return to traditional, physical classroom environments seems unlikely, if not impossible. The Covid-19 pandemic has merely accelerated what was already a significant trend in education: the merging of virtual and physical learning spaces. Although the exploration of this blended reality began long before the pandemic, the urgency of recent events has made it a real task for educators, students, and technology developers alike. In an early look at this trend, Bower et al. (2010) explored the possibilities and challenges of combining virtual reality with face-to-face communication in a classroom setting. They defined blended reality as a closely related physical and virtual environment that works together to achieve specific educational goals. Through the lens of activity theory, the authors evaluated an experimental setup where both realities were augmented. The findings revealed the feasibility of simultaneous interaction with physical and virtual participants, though there were complications related to video, audio quality, and camera angles.

Similarly, Chen et al. (2010) undertook a study to compare the effectiveness of direct and interactive instructional styles in both virtual (specifically in Second Life) and physical learning environments. They found that interactivity was key, with the interactive approach yielding greater student involvement in virtual environments. However, they also discovered that more cues were needed in a virtual environment, and that direct instructional style was less useful there. Holmberg and Huvila (2008) investigated the comparison between Second Life and traditional face-to-face environments in distance education. They concluded that Second Life could replicate the experience of in-person education, even benefiting teamwork and reducing the issues of distance in learning. This is in part due to Second Life's multimodal communication features, allowing for smooth, varied communication. Wuensch et al. (2008) broadened the scope to 46 American universities, studying the differences in pedagogical features between in-person and online classes. While online courses were seen as more convenient, they lagged behind in terms of communication quality and understanding of course material. The authors emphasized the need for technological improvements and better consideration of the needs of both teachers and students. Together, these studies illustrate a complex and evolving landscape. The blending of the virtual and physical worlds offers exciting opportunities but also brings significant challenges. It's not simply a matter of choosing one approach over the other; it's about finding the right balance. We can't go back to purely physical classrooms, but we also can't ignore the unique characteristics and requirements of virtual environments. As educators, technologists, and learners continue to navigate this new terrain, it will be crucial to learn from these early studies and others like them. The goal must be to create hybrid learning experiences that are engaging, effective, and satisfying for everyone involved.

VR and AR Tools

In their 2020 study, Nesenbergs et al. conducted a literature review of nine studies, examining the effects of VR and AR on learning outcomes in higher education during the Covid-19 pandemic. Their findings highlight the potential of VR and AR for practical and laboratory classes when physical attendance isn't possible. However, they caution that these technologies cannot entirely replace in-person learning, as it might negatively affect student grades. Out of the thirty AR technologies featured, eleven were found to positively impact performance and six enhanced engagement. The use of AR and VR was found to facilitate students' understanding of abstract, complex concepts, with kinesthetic learning in the virtual world proving more effective than in traditional environments. Nesenbergs et al. suggest that AR and VR could almost mirror real-world learning environments, making successful global training via these platforms a possibility, especially relevant during the Covid-19 pandemic, where online learning is seen as a pathway to future education. Similarly, Pregowska et al. (2021) undertook a comprehensive literature review, delving into the history of distance learning across various regions, as well as the roles of AR, VR, and mixed reality in distance education, contextualized within the Covid-19 era. Their study spans more than 100 publications dating from 1926 to 2021. Starting with the early beginnings of distance education in America, Europe, Australia, Africa, and Asia, they trace the evolution of mediums from radio and television to CDs and the internet. The focus then shifts to current applications of VR, AR, and mixed reality, especially in science and medicine. Following the onset of the Covid-19 pandemic, there has been a swift global pivot toward distance education as a containment measure. Pregowska et al. explore how different countries have adapted and discuss the advantages of distance learning, such as increased flexibility and accessibility. Conversely, they also highlight the drawbacks, including limited internet access and the absence of personal feedback and physical interaction. This comprehensive review adds to the discourse on the importance of hybrid education models in contemporary times, emphasizing that the journey toward the right balance between virtual and face-to-face learning is an ongoing and crucial aspect of our global educational landscape.

Face-to-Face Versus VR with an Avatar Versus VR Without an Avatar

In their 2018 study, Smith and Neff investigated communication behavior within "embodied virtual reality," where user movements are imposed on a virtual avatar. The study aimed to compare three types of communication: face-to-face, VR with an avatar representing the user, and VR without avatars. They conducted experiments with thirty pairs of participants, engaging in role-play tasks within the VR environment. Through detailed analysis of gestures and conversational turns, they found that verbal and nonverbal communication styles were generally similar in face-to-face interaction and when an embodied avatar was present. The absence of the avatar, however, led to decreased referential pronoun usage and conversational turn frequency, suggesting that having a tracked body in the virtual environment enhanced the feeling of interacting with another person. This study's insights contribute to a broader understanding of communication in the metaverse, particularly in applications like public speaking training or reducing public speaking anxiety.

From Virtual Therapy to Enhanced Feedback: Exploring VR and AR in Public Speaking

The landscape of public speaking training is evolving with the integration of VR and AR technologies. Studies by Davis et al. (2020) and Damio and Ibrahim (2019) have begun to unravel the potential of VR as a practice tool for public speaking. Although students experienced greater anxiety in virtual sessions, the outcomes were found to be comparable to traditional practice methods. In examining the nuances of VR and AR, Ling et al. (2012) questioned the practical relevance of stereoscopy, while Parmar and Bickmore (2020) highlighted the usefulness of 3D spatial overlay visualization in providing feedback to speakers. Poeschl (2017) introduced a framework to evaluate the quality and effectiveness of VR training applications, revealing insights into how concentration, fear, and task difficulty impact performance. Social phobias and the fear of public speaking were a central focus for several researchers. Slater et al. (1999) discovered that virtual environments could be therapeutic, with audience interest and co-presence playing pivotal roles in shaping emotional responses. Similarly, Takac et al. (2019) showed that successive speaking scenarios in VR could elicit distress, whereas von Ebers et al. utilized wearable devices to predict the effectiveness of VR in reducing anxiety. An interest in enhancing the public speaking experience through technology was also evident. Yadav et al. (2019) explored how systematic VR exposure could reduce anxiety levels, while Zarraonandia et al. (2014) leveraged AR to improve communication during presentations. The intersection of VR and AR with public speaking offers a fascinating glimpse into the future of communication training. These studies collectively highlight promising avenues but also underline the complexity and variability of these emerging technologies. The effectiveness of VR and AR in this context continues to be an area ripe for further exploration and refinement.

7

Case Studies

This chapter delves into a specific case study that my team conducted in the post-Covid era of 2022 to 2023, where metaverse platforms were utilized to teach courses in Korean language, intercultural communication, and digital humanities. Drawing from my personal teaching experiences during this period, the study amalgamates observations and insights garnered from interviews with fellow educators on my team. As emphasized in Chapter 6, while individual case studies should be carefully assessed without broad generalizations, they can serve as valuable pointers. Accordingly, this presentation of my experience is intended as a contribution to the evolving dialogue on education in the metaverse, albeit with the understanding that its applicability will be limited to similar contexts.

Background and Methodology

In the realm of language education, innovation often serves as a catalyst for enhanced learning experiences. This was the guiding principle when I embarked on a unique instructional journey in the domain of Korean language teaching. Over sixteen weeks, in collaboration with my team, I designed and implemented a program centered around Gather Town, a 2D virtual platform. This initiative was driven by a desire to explore how virtual platforms could complement and enhance the traditional teaching methodologies inherent in physical classrooms. Our student cohort experienced a combined learning structure: They engaged in interactive sessions within the virtual landscape of Gather Town and concurrently attended conventional classes in physical settings. The metaverse component wasn't a mere supplementary tool; it constituted a significant part of our teaching strategy. Within Gather Town, we designed 50-minute sessions meticulously crafted to engage students in exercises and activities aimed at refining their Korean language skills. By merging the immersive and interactive attributes of the metaverse with the direct, human-centric advantages of face-to-face instruction, our goal was to offer students a holistic learning experience. This combined approach was an experiment, one which sought to capitalize on the strengths of both virtual and real-world instructional methods. In the early stages of integrating the metaverse into my teaching approach, I was enticed by the potential of 3D platforms. Their visually rich environments seemed ideal for crafting a deeply immersive learning experience. Yet, as we delved deeper, the practical limitations of this technology became evident. 3D platforms, despite their appeal, come with significant challenges: the technology's complexity, the associated costs, and the discomfort it poses for individuals with glasses or visual impairments. Notably, many users found that they could comfortably use 3D headsets for only about 10–15 minutes at a stretch, making them unsuitable for extended class sessions. 2D spaces, in contrast to VR headsets, present several benefits. They allow for longer, uninterrupted use, making them more conducive to sustained learning sessions. Additionally, they are generally more accommodating for glasses wearers. Based on our assessments, we concluded that, for the present, VR headsets may not be the most effective tool for language learners. Given these challenges, our team chose 2D metaverse platforms, selecting Gather Town as our primary instructional tool.

Detailed Observations from a Dual Learning Environment

Over sixteen weeks, my colleagues and I set out on a structured experiment that bridged both traditional and digital educational methodologies. Tasked with the challenge of teaching Korean in this blended environment, we realized the scarcity of established guidelines or studies on how best to integrate the physical and virtual, especially in the domain of language teaching in the present digital landscape. We structured our teaching approach around a dual classroom model. This involved making use of physical classroom spaces, which offer the undeniable benefits of face-to-face interaction, while also leveraging the virtual environment provided by Gather Town. The motivation behind this hybrid methodology was not merely to adapt to current trends but to understand, in-depth, how each mode of teaching can enhance the other. Throughout this period, our roles evolved beyond mere instruction. We were active observers, carefully documenting the varied nuances that emerged from our experiment. The challenges we encountered were as valuable as our successes, each offering a unique perspective on how digital and traditional methodologies interact. Our primary goal was to offer students an enriched learning experience, balancing the direct engagement of physical classrooms with the expansive, interactive potential of the metaverse.

Time-Saving Benefits

Our students prefer the metaverse because of transport and travel time. Many have to take the bus to reach the centre of the city. Sometimes language learning is just a hobby, so time is important.

This quotation is an observation from one of the teachers of the Korean language sessions. Most of the participants in the Korean courses were not centrally located in Oxford. In fact, many resided outside Oxford, some as far away as London. This posed challenges, especially during the winter. The biting cold made commuting difficult for our participants, of whom a significant number were either employed in day jobs or engrossed in pursuing their degrees. While juggling these responsibilities, finding time to study was already a challenge. But to add the demands of travel to this equation made consistent attendance even more elusive. This is where the metaverse offered a significant advantage. It provided these learners with a convenient avenue to continue their education. Participants could engage with the course without the added strain of commuting. However, what set this experience apart from simply watching a YouTube video or streaming other lectures was the immersive and interactive nature of the metaverse.

Recent findings from a global survey by the *Times Higher Education* have unveiled a discernible trend. Post-pandemic, student attendance and participation have seen a considerable decline. Even with Covid-19 restrictions lifted, a striking 76 percent of academics reported a drop in student attendance (Penn, 2022). In stark contrast, a mere 4 percent witnessed a rise (Penn, 2022). Further adding to the concern, over half of these educators observed decreased student engagement during lectures. The reasons behind this trend are manifold: reluctance among students to return to campus, the allure of paid work, struggles with mental health, and even a lack of preparation for lectures. As these challenges loom large, educators and institutions are confronted with a pressing question: How can the classroom experience be revitalized? I think the answer lies in finding a way to balance physical and metaverse classrooms. The metaverse classroom presents a compelling solution to one of students' primary concerns: time. The traditional classroom experience, while valuable, comes with the often underacknowledged cost of commuting. Be it a short trek across a vast campus or a draining journey across the city, commuting consumes both time and energy. However, the metaverse classroom reimagines this dynamic. The virtual setting obviates the need for physical commutation. Students, armed with a stable internet connection, can now attend lectures from any corner of the globe. Gone are the logistical challenges of planning around weather, traffic, or even the daily packing of essentials. This shift goes beyond merely saving hours; it enhances the overall quality of life for students. Freed from daily commuting fatigue, students can approach their studies with renewed vigor and enthusiasm.

Embracing this digital transformation could be the revitalization academia seeks. The blending of the traditional with the digital highlights a pivotal moment in education. It underscores the need for adaptable teaching methods that cater to the evolving preferences of students, maximizing their time and potential. As we progress, it's clear that the academic realm must reconceptualize "attendance" and teaching methodologies for a digital age. It's worth noting that metaverse platforms are not just alternatives but likely represent the future of education. Given the shift in the culture of attendance and the persistent timing constraints students face, there's a clear inclination toward hybrid models that combine the physical and the digital, with metaverse platforms taking center stage.

Psychological Comfort

In our metaverse classroom, I've observed a noticeable shift in the learning dynamics of my students who are studying Korean. The virtual setting seems to offer an environment that encourages them to express themselves more openly. One clear difference I've noticed is in the frequency and nature of their questions. In a traditional classroom, students might hesitate to ask questions they fear might be perceived as "silly" or "unconventional." However, in the metaverse, there's an increase in such inquiries. This suggests not only heightened participation but also a greater comfort level in their learning space. It's particularly intriguing that some students, who were typically reserved in the physical classroom, have become quite active in the metaverse environment. They often send multiple queries through the chatroom, showcasing a newfound enthusiasm and engagement. This change could be due to the perceived anonymity the virtual space provides or the lack of immediate peer judgment that can be present in a face-to-face setting. One instructor in my team said:

> We have a student who will never give an answer in physical class, but on the metaverse he has more confidence because he knows he doesn't need to have his camera on or use his voice. He can put his answer in the chat box. Some students have a bad relationship with teachers who have said to them, "What is the answer? Get it right!" The metaverse reduces that pressure. I want them to enjoy it. They won't learn properly if they're not enjoying it.

> In a breakout area only one small group of students interact with each other. The breakout areas are also designed so that the teacher can listen in without students knowing. This is great because they don't feel pressure from knowing that we are listening.

While the metaverse classroom has its advantages, like creating an inclusive environment where more students feel at ease to participate, it does have its limitations. In a conventional classroom, the flow of discussion is spontaneous, with immediate reactions and interactions. Although time constraints can limit some students from asking questions, the opportunity for real-time, in-person engagement is something the metaverse environment hasn't fully captured.

Often, in the physical classroom setting, there has always been a distinct divide in student engagement. A subset of students consistently participates, asking questions and driving the discussion, while others often remain quiet, blending into the background. This has been a longstanding dynamic, where the more extroverted or confident individuals tend to dominate the discourse. However, in the metaverse classroom, this dynamic seems to be shifting. The playing field appears to be leveling out, with more students engaging openly and equally. Those who typically reserve their questions or comments in the physical classroom are finding their voice in the virtual environment. The metaverse seems to offer a space where the usual barriers or inhibitions that deter some students from participating are reduced. The balance of engagement is more evenly distributed among the class, allowing for a more diverse range of insights and perspectives. Combining this with the earlier points, it becomes evident that while the physical classroom allows for immediate, spontaneous interactions, it might inadvertently sideline some students. In contrast, the metaverse environment, despite its own set of challenges, creates an inclusive space where students who might have otherwise remained silent feel empowered to participate actively. Recognizing and integrating the strengths of both settings can lead to a richer, more inclusive educational experience for all students.

Sense of Presence: Visual Engagement

In the metaverse classroom setting, students found themselves engaging in preparatory actions before the actual session began. There was an undeniable sense of presence and mutual awareness, as everyone could view each other's avatars. This dynamic is highlighted in an interview extract with one of the students, emphasizing the tangible feel of the virtual space and the consciousness of being part of a collective learning experience:

> We set the metaverse so that there is a foyer upon entrance. Students first enter an Oxford style common room with sofas, armchairs, a fireplace, and 4 private areas to talk. Then, to start the lesson they walk through the door to a new space. In the foyer, they can check their camera and mic. They can prepare. It breaks the ice every time they come into the lesson. This is very beneficial for having them focus for an hour.

From our experiences, we have seen first-hand how vital visual engagement is to fostering genuine connections:

> As a teacher, I feel more comfortable with the avatars. With Zoom, having your camera off and mic muted is very impersonal, but with avatars, it feels like my students are still there. This keeps the teacher's motivation up. This is especially useful for sign up talks where you don't know who anyone is, and you don't know who will turn up. Talking to a list of names can feel like talking to robots. The metaverse provides a physical representation of how full that room would be in real life.

On platforms like Zoom, it's often challenging to get a comprehensive feel for who's present in a session. The gallery view provides little thumbnails of attendees, but with cameras turned off or nondistinct backgrounds, the sense of presence is often diminished. It can feel like communicating into a void, with minimal cues on attendees' engagement or attention. Transitioning to the metaverse, especially in platforms like Gather Town, has shifted this dynamic considerably. Here, the use of avatars has brought about a more tangible sense of presence. For instance, before a class begins, I can observe students' avatars taking their seats, signifying their readiness to engage. This simple action, akin to students settling into their chairs in a physical classroom, offers a powerful visual affirmation that they are present, both in attendance and in focus. Avatars in Gather Town and similar platforms do more than just mark attendance. They foster visual connections among students. Seeing avatars move, congregate, or even gesture can provide cues about group dynamics, levels of interest, and even spur spontaneous interactions, much like how body language works in the real world. These visual cues serve as catalysts for enhancing interaction, allowing both educators and students to read the virtual room and adapt accordingly. The ability to modify and adapt one's virtual presence can alleviate some of the pressures or anxieties students might feel about their physical appearance in real-life settings. In the metaverse, it's less about conforming to societal expectations and more about individual expression and comfort, providing a unique way for students to connect with their learning environment.

Additionally, in the metaverse, the concept of physical appearance takes on a unique dimension. Unlike traditional classrooms, where students might feel conscious about their attire, hairstyle, or general appearance, the virtual realm offers a distinct kind of flexibility. For instance, a student mentioned:

> *There is a physical sense within the virtual space. Logging on and thinking 'what colour hair shall I have today?' prepares you for class and gets you ready to learn.*

This reflects how the metaverse allows for a reimagining of one's identity and presentation on a day-to-day basis. It's not just about customizing an avatar—it's about expressing oneself, experimenting, and even using these choices as a preparatory step to mentally engage with the upcoming session.

Sense of Community

> *The metaverse feels so much warmer than Zoom. The aesthetic design is nostalgic because of the 2D and the avatars. Seeing other avatars makes you feel connected without the pressure of physical lessons. If you have two or three lessons every week, and someone asks for help, you can easily meet them in a virtual classroom, and you can go back to the class materials saved there very easily.*

This is another observation from the Korean sessions. The advent of the 2D metaverse classroom has proven transformative for community building within our teaching methodology. We discerned a marked shift in the ways students forged connections and interacted among themselves. From the outset, it became abundantly clear how vital fostering a communal sense was to the overall educational experience. When students identify with and feel integral to a collective, they are naturally inclined to be more participative, collaborative, and wholly immersed in their learning. Transitioning to online modalities had its initial hurdles, primarily the diminished sense of community. The lack of tangible presence, the void of physical gestures, expressions, and the spontaneous, organic interactions one finds in brick-and-mortar classrooms led to an attenuated learning atmosphere. However, 2D platforms like Gather Town heralded a new dawn in virtual education, addressing these very challenges.

In these virtual spaces, students weren't merely passive observers of pixelated avatars. They experienced a profound sense of being and belonging. Observing avatars move, interact, or simply position themselves in their designated seats imparted a sensation of closeness and connectedness rarely achieved in traditional online platforms. This newfound virtual proximity was not just physical but emotional and psychological, instilling a genuine feeling of togetherness. As students navigated these spaces, they began to feel more connected, not just to the content or the educator, but to each other. This environment was reminiscent of a traditional classroom, where unplanned discussions, off-topic chats, and spontaneous debates emerge, fostering more than just academic growth. They shared laughter, experiences, and stories, building relationships that transcended the confines of the virtual world. This intrinsic sense of community and the resultant feeling of connection led to not just enhanced academic engagement but also the nurturing of bonds and mutual respect among students. In essence, the 2D metaverse classroom did not just replicate the physical classroom experience but enriched it, reminding everyone that at the heart of education lies the human connection.

Flexible Resources

One of the key practical benefits we discovered in our course was the enhanced capability for sharing resources within the metaverse, specifically in Gather Town. Gather Town's capabilities addressed and solved many of the limitations we faced on traditional platforms, making resource sharing both seamless and effective. This flexibility has undeniably elevated the quality of our teaching and the depth of our students' learning experiences. For our students learning Korean, motivation often stems from a deep appreciation of Korean culture—be it drama, K-pop, or films. Thus, the integration of multimodal materials into our teaching approach became crucial.

In traditional online teaching platforms, like Zoom, we consistently faced challenges when attempting to share these rich, multimedia resources. Technical glitches, file size limitations, and the limitations of real-time sharing often hampered our efforts. It wasn't just about the content; the very act of sharing became an obstacle. However, the digital space of Gather Town improved this aspect of our teaching. The platform enabled us to upload a diverse range of materials, from videos to presentations, without the usual restrictions. Not only did this mean that students could access resources during the lesson, but they could also revisit them at their convenience, further enhancing their learning experience. Moreover, when it came to materials like PowerPoint presentations, the metaverse allowed students to have individual control. They could review, pause, and reflect on slides at their own pace, empowering them to take charge of their learning journey.

In the initial phases of our case study, there were certain outcomes I hadn't anticipated. One such revelation was aptly summarized by a teacher:

Students can access the classroom whenever they want. They can use the space for a study session; they don't have to book a room in the library.

This statement opened my eyes to an unexplored facet of the metaverse in an educational context. Rather than just a digital classroom, the metaverse emerged as a dynamic space that blended the utility of a library with the social ambiance of a café. Students weren't merely attending classes; they were turning these virtual spaces into social hubs, places where learning coexisted seamlessly with interaction. This fluidity between study and socialization is a dynamic I hadn't fully appreciated until the study was underway.

Uninterrupted Collaboration

One of the standout benefits of transitioning to the metaverse classroom was the remarkable ease and efficiency with which pair work could be conducted. Pair work in language learning settings is important but achieving this seamlessly on platforms like Zoom presented numerous challenges. On video call platforms, every spoken word is broadcasted to the entire group, making intimate side conversations or small talk a challenge. The "breakout room" feature, intended for facilitating smaller group interactions, often proved limiting, preventing teachers from effectively monitoring and giving real-time feedback. The metaverse, especially on platforms like Gather Town, addressed these issues head-on. The platform's spatial audio feature ensures students only hear someone when they're close to their avatar. This not only allows for genuine one-on-one discussions but also replicates the real-world dynamics of classroom interactions, where two students can have a private conversation without the entire class eavesdropping. Additionally, the spatial arrangement of the platform empowers teachers. They can "walk" through the virtual space, drop into pair discussions, and provide guidance, just as they might in a physical setting. This interactive capability ensures that pair activities remain both productive and under the instructor's purview.

Enhanced Interaction in a Nonhierarchical Environment

During my time teaching in the metaverse, I noticed a marked uptick in cross-group interactions. Students who might not usually engage with each other in a conventional classroom found common ground in this new setting. These interactions weren't just cursory; they were meaningful exchanges, ranging from academic collaborations to cultural exchanges and shared digital experiences. This amplified interaction is invaluable, especially in an educational context. When students interact with peers from diverse backgrounds and perspectives, it enriches their understanding, promotes empathy, and fosters a more inclusive learning environment. It's akin to opening windows to different worlds, all within a single virtual classroom. From my personal experiences within the metaverse classroom, one of the most noteworthy observations was the shift in the traditional teacher–student dynamics. In this digital space, everyone, whether a teacher or a student, is represented by an avatar. This uniformity in representation goes beyond mere aesthetics; it plays a pivotal role in shaping interactions and establishing equal dialogues.

The metaverse provides a unique environment that can reshape longstanding educational norms, promoting inclusivity and dialogue over authority and hierarchy. This shift not only lessens student anxiety but also lays the foundation for a more progressive and collaborative learning experience. The similarity in the appearance of avatars means there's no immediate visual cue differentiating a teacher from a student. This subtle equalizer can significantly lower the psychological barriers that often exist in more traditional educational settings. Without the usual indicators of authority or rank, conversations become more open, and the environment feels less intimidating for students:

> *Teacher student dynamics feel more equal in the metaverse. The set up in our physical classroom has the whiteboard on one side and screen on the other. Students sit around the table while I stand up. That feels stricter. Online, I wear casual clothes, in the classroom, I wear formal clothes. So, I feel more comfortable. Overall, it's less hierarchical. I'm younger than most of our students, so to stand in front of them as a non-native speaker could be quite stressful. I don't want them to think*

that I am overly strict and bossy. In the metaverse, I am sat down, so it is easier on a personal level. I feel like I can make a mistake and it won't be the end of the world.

This nonhierarchical atmosphere is particularly beneficial when considering the cultural context of education in regions like Asia, where educational dynamics often lean heavily on authority and respect for elders. In such settings, the inherent hierarchy can sometimes stifle open dialogue, discourage students from expressing dissenting opinions, or asking questions for fear of seeming disrespectful or ignorant. In the metaverse, this balance is redressed. The comfortable, open setting encourages a more relaxed teacher–student dynamic. Students can voice their opinions, ask questions, and interact with both peers and educators without the weight of hierarchy influencing their actions. The result is an enriched classroom dynamic where engagement is genuine, discussions are candid, and learning becomes a collaborative process.

A More Inclusive Classroom

When it comes to fostering a robust educational atmosphere, cultivating a sense of social belonging and inclusivity is paramount. Avatar-based platforms, like those in certain metaverse settings, can provide a middle ground. They ensure active participation and create a sense of community without inadvertently adding to the stressors students might experience. It's a way of ensuring that the educational environment is as inclusive and accommodating as possible, respecting both the need for social interaction and individual comfort level. It's not just about knowledge transfer; it's about creating a space where students feel seen, heard, and valued. However, achieving this inclusivity doesn't necessarily mean that everyone should be completely open or make themselves fully visible. There's a delicate balance to strike. In traditional classroom settings or even in typical online platforms, there's an implicit expectation for students to have their cameras on all the time. This continuous visibility can sometimes inadvertently create extra pressure. Students might feel the need to present themselves in a certain way or might be uncomfortable being in the spotlight continuously. Using avatars, students can be present, engage in discussions, and contribute to the learning environment without the added pressure of being visibly on display. It offers a form of representation that is both social and semi-anonymous. This way, students can engage genuinely without the potential anxieties or burdens of constant visibility:

> Students have so much pressure in their lives already, if they need to turn off their camera, it's okay. If they haven't tidied their bedroom, they're in a public place. or they share a bedroom, then the metaverse avoids you being limited by your environment. You can still engage with the lesson without worrying what people think about your background, while avatars provide a sense that you are learning alongside classmates. With Zoom, if you don't turn on the camera you look antisocial, but in the metaverse, you're still there, and people know who you are.

Engaging Gamification

Teaching in the metaverse opened doors to many possibilities, one of which was the power of gamification. The integration of gamified elements in my class showcased the potential of the metaverse as a dynamic learning environment. When students are entertained and engaged, the boundaries of traditional learning expand, leading to more enriched and memorable educational experiences. The nature of language learning, while deeply rewarding, can sometimes be challenging and tedious. Hence, finding ways to infuse elements of joy and fun into the process can be instrumental. One memorable experiment of mine in this space was introducing virtual motorbikes to the digital classroom. I placed five digital motorbikes within the metaverse environment, curious to see how my students would react. The outcome was beyond my expectations. Students started coming early to class, eager to find these virtual motorbikes and ride them around before the session began. This simple gamified element not only added a playful touch but also managed to instill a sense of anticipation and excitement for each class. However, the surprises didn't stop there. As the students grew more comfortable and engaged within this gamified environment, they began to voice their ideas and suggestions. Some came forward with creative ideas for decorating the space, while others recommended games or interactive elements that could be incorporated into our sessions. This feedback loop not only made the classroom more collaborative but also ensured that the space was continuously evolving and tailored to the preferences and suggestions of its primary users—the students.

Beyond the motorbikes, other gamification elements like shared keyboards and avatar decorations amplified this interactive and fun atmosphere. Avatar decorations, for instance, provided students an opportunity to showcase their personalities, thus creating a personal connection within the digital space. The beauty of these gamified features was that while students were engaging with them, they were also subconsciously practicing language skills. Chatting about the motorbikes, decorating avatars, or playing games invariably involved using the language they were learning, thus reinforcing their skills in a relaxed environment.

Socialization Opportunities

The metaverse, especially platforms like Gather Town, offers more than just a digital replica of a classroom. It can present a holistic environment that captures the essence of a physical learning space, right down to the foyers and common areas that are integral to student interactions. In my experience with the metaverse classroom, I discovered an unexpected yet delightful phenomenon: Students were actively socializing outside of scheduled class times. Much like the spontaneous gatherings one might witness in the hallways or common areas of a traditional school or university, students in the metaverse were congregating in the virtual foyers and open spaces, indulging in conversations, games, and shared experiences.

This behavior underscores the profound need for social interactions in the learning process. While classroom time is undeniably valuable, the informal conversations and interactions that occur outside the classroom are equally crucial. They provide students an avenue to share their thoughts, clarify doubts, and build relationships in a less structured setting. In Gather Town, these social spaces become hubs of continuous learning. It's not just about academic discussions or group study sessions. Students share personal anecdotes, cultural insights, and even casual banter, all of which contribute to a richer understanding of the subject at hand, particularly in the context of language learning, where cultural nuances and colloquialisms are paramount. What stood out was the realization that learning wasn't confined to classroom walls, be they physical or virtual. By simulating foyers and other informal spaces, Gather Town has inadvertently nurtured an environment where learning permeates every nook and cranny. Students, drawn by the allure of these social spaces, are motivated to continue their educational journey even outside of structured class hours, emphasizing that learning is truly a continuous process.

Teaching New Scripts in the Metaverse: An Advantage in Language Education

In the realm of language teaching, particularly for languages with distinct scripts, the challenge of ensuring students grasp the nuances of writing can be daunting.

Korean, with its Hangul script, is a prime example, but this extends to other languages across Asia and beyond, such as Chinese with its intricate logograms, Japanese with its combination of Kanji and Kana, or Hindi with its Devanagari script, among others. During our Korean teaching sessions in the metaverse, we noticed a significant advantage in using the platform's interactive features for teaching scripts. One of the teachers in our team remarked:

> *Gather Town provides interactive features like whiteboards that allow students to practice writing. This not only enables us to monitor their strokes and actions in real-time but also offers a distinct advantage over more traditional methods. For instance, when students write in their personal notebooks out of a camera's view, we might miss out on their mistakes or nuances in their writing. With Gather's features, immediate feedback can be provided, and saved work allows for extended review and correction outside immediate sessions.*

The ability to see, correct, and provide real-time feedback on students' writing becomes exponentially valuable when teaching languages with unique scripts. This visual element, combined with the saved work feature, not only promotes better comprehension but also caters to the iterative nature of learning scripts. Students can repeatedly practice, make errors, correct themselves, and gain confidence in their writing abilities, all under the watchful eyes of their educators. While our case study revolved around teaching Korean, the implications of these findings could be broadened. Educators tasked with teaching languages that involve learning new scripts could significantly benefit from leveraging such metaverse platforms, ensuring that students gain a strong foundation in their writing skills.

Learners as Teachers

An important lesson I personally learned from teaching in a metaverse classroom is that I'm not necessarily instructing the students or superior to them in any way—especially when it comes to digital literacy. In fact, the metaverse classroom enables a vibrant culture of peer learning. As our classrooms increasingly migrate to the digital world, this dynamic is likely to become even more prevalent. Younger generations, already adept at navigating digital spaces, will find themselves in roles where they can teach adults, including their teachers. This shift is not just confined to digital literacy but extends to content generation as well. Students can actively participate in shaping the curriculum based on their interests, thereby enhancing their engagement and the relevance of what they learn. In this digitally evolved setting, teachers serve more as a guide, introducing new methodologies and allowing students to immerse themselves and explore on their own terms. The role of a teacher transforms from a primary deliverer of information to an inspirer, overseer, and provider of constructive feedback. The teacher's job is no longer to deliver the entirety of learning, but to facilitate a space where learning can occur in a more organic, student-led manner.

The empowering dynamic of the metaverse classroom disrupts the traditional teacher–student hierarchy. While typical classroom dynamics can be anxiety-inducing, a well-moderated metaverse environment can mitigate this, creating a more egalitarian learning experience. In this space, not only can students become teachers, but everyone can also learn together. This cultivates an ideal environment where learners can truly take ownership of their educational journey, thereby making the learning outcomes more impactful and long-lasting. Given these emerging trends, it's clear that the educational landscape is undergoing a significant transformation. The transition to digital classrooms and the shift in roles between teachers and students underscore the necessity for adaptive, responsive educational models that empower learners as active participants in their own education.

Another Case Study: Teaching Intercultural Communication

In the following discussion, I shall delineate findings from a case study focused on the potential of the metaverse for instructing intercultural communication. The importance of intercultural communication in our interconnected world cannot be overstated. Yet, facilitating large, diverse groups for substantive interactions presents significant logistical challenges. For the study, I set up four workshops centered on the shared topic of their daily food and drink. These workshops incorporated participants from Hong Kong, the Czech Republic, and the UK. The feedback obtained highlighted several key observations. A notable proportion of participants expressed increased comfort within the metaverse, citing its conducive environment for learning. The platform seemed particularly apt for smaller, focused discussions, facilitating a deeper exploration of cultural variances. We engaged in multimodal activities, watching video clips as if in a virtual cinema. While some participants encountered initial adaptability challenges, the overarching sentiment was positive, with many acknowledging the metaverse as a beneficial tool. Based on the accumulated data and my experience, the metaverse exhibits promising utility for intercultural communication instruction. After the workshops, we collected feedback from the students. I will now discuss some observational notes based on this feedback.

More Fun and Engagement

First and foremost, the majority of feedback we garnered leaned toward the positive side. Participants often characterized the experience as having a playful essence, with reduced feelings of anxiety. It's particularly noteworthy that, despite most participants not being native English speakers, they actively engaged in English communication. This pattern of feedback mirrors the insights I previously gathered from teacher interviews during our Korean workshops.

> *I felt more confident in expressing my ideas in English in the metaverse. I was less stressed. Because I was staying in a room where only my group members could hear me, that made me feel less anxious to speak up. Besides, the whole setting, such as the characters and the venue, is cute which makes me relax and feel more engaged.*
>
> *It was a really enjoyable experience. At first, I was a bit anxious, but then the anxiety disappeared.*
>
> *In my view, the metaverse fosters a relaxed environment for open discussions. Turning off my video eased my anxiety, and the supportive group members encouraged active participation. While I still favour face-to-face conversations, metaverse interactions come a close second. I genuinely enjoyed the experience.*
>
> *Meeting new people this way is fun and innovative, albeit slightly awkward at times.*

Both participants from the Czech Republic and Hong Kong expressed feeling more at ease speaking English in the metaverse. They attributed this comfort to being able to engage in small-group discussions. Additionally, the aesthetic elements of the metaverse made discussions feel more approachable and less intimidating. However, it's worth noting that if serious topics are to be taught in the metaverse, the classroom setting should ideally reflect the gravity of the subject to avoid any potential undermining of the content.

More Expressive

A few participants also highlighted the expressive potential of communication within the metaverse. They particularly emphasized the positive impact of using avatars. Additionally, they noted the benefits of the metaverse in terms of global educational connections, underscoring the convenience of not needing to travel.

> The metaverse allows for unique self-expression through avatars, which I find more exciting than a mundane profile picture. It's a fantastic medium for cultural exchange and learning, especially for those unable to travel globally. Being part of this diverse and creative community has been both fun and enlightening.

Psychological Stability: AI-assisted Versus Traditional Approaches

While more in-depth studies are required, based on my initial pilot studies, it appears that a significant distinction emerges concerning the psychological aspect of learning between traditional methods and AI-assisted learning. In the traditional learning landscape, the learning process can be likened to a rollercoaster ride of emotions. As learners progress through text or course materials, they often experience a range of emotions, including frustration, excitement, confusion, and satisfaction. These emotional fluctuations sometimes lead to a sense of instability in the learning journey.

In contrast, AI-assisted learning introduces a distinct dynamic. By harnessing technology, machine-learning algorithms, and personalization, AI strives to provide a more consistent and steady learning experience. Its goal is to engage students in a manner that minimizes emotional turbulence. One notable advantage of AI-assisted learning is its adaptability to individual learning styles and paces. AI can customize content and assessments to align with each student's unique needs, thereby creating a psychologically stable learning environment. This personalization can mitigate the emotional highs and lows that learners may encounter in traditional educational settings.

More to Follow-Up On

While there's great potential in the metaverse and it presents an exciting space for education, it's essential to proceed with caution and further investigation.

Notably, not everyone finds the environment anxiety-free; for some, it may heighten anxiety levels. Technical glitches are always a potential issue, and interacting with someone new in this setting can be daunting. There's much more to explore through hands-on practice. However, based on my initial investigations, areas such as language instruction, global studies, and media studies could greatly benefit from this platform. For instance, some participants said that it could work better in small groups:

> *The metaverse is better suited for interaction in small groups rather than large ones. While it allows us to meet, speaking to an unknown avatar initially felt a bit daunting.*

This observation presented an intriguing duality. Initially, there was a sense of anxiety, especially because you may meet someone you've never met before without seeing their face. Various factors contribute to this anxiety, and such feelings are only natural. However, this anxiety tends to resolve after a series of interactions. While the metaverse has much to offer, there's an equal measure of exploration and understanding required. Areas like language teaching, global studies, and media studies stand to benefit immensely, but the journey will necessitate ongoing research and adaptive strategies.

Annoyance and Boredom: A 2023 Perspective on Non-English Languages and Beyond

In my own case studies, I observed that learners often feel annoyed when engaging with current AI technologies, particularly in non-English languages. In the context of 2023, AI systems can frequently make mistakes or misinterpret queries, which can be particularly frustrating for users. Furthermore, while the initial interactions with computer-assisted learning may elicit a sense of wonder, it is not uncommon for users to experience moments of boredom once the novelty wears off, especially when compared to the excitement of gaming.

Metaverse in Education: A Call for Comprehensive Exploration

The swift rise of the metaverse in educational contexts illuminates both a unique opportunity and a challenge, underscoring the immediate need for robust, comprehensive studies in this arena. Given the freshness and intricacy of metaverse learning environments, there's a palpable gap in empirical research assessing their impact across diverse pedagogical scenarios. While initial observations, such as mine, provide beneficial insights, it becomes essential to stretch these explorations to encompass varied age groups, subjects, and teaching techniques. The journey of a high school student traversing historical landscapes might differ markedly from that of a university scholar delving into media studies or an adult learner navigating the corridors of a new language. Furthermore, the perspectives of educators and learners, while sometimes overlapping, can offer distinct insights, each playing a pivotal role in shaping our understanding of metaverse education. Relying excessively on limited data or making hasty generalizations can pave the way for inadequate or misaligned educational strategies. In light of this, there emerges a compelling case for interdisciplinary, extensive research endeavors, aiming to mold the vast potential of the metaverse to fit the nuanced needs of different learning ecosystems.

8

Future of the Metaverse

The metaverse isn't just a futuristic concept anymore; it's now a defining aspect of our current reality, actively molding what our future looks like. As we've discussed throughout this book, our challenge is to figure out how to seamlessly merge the physical with the digital. This demands a shift from our conventional paradigms, as we account for the new dynamics and opportunities the metaverse offers. It's not just about replicating our physical world digitally; it's about forging a new domain where the digital and physical realms synergize, each enhancing the other. Venturing into this uncharted territory requires careful crafting and thoughtful decision-making.

The metaverse is more than just a novel alternative space, as it's also a potential inevitability we might find ourselves embracing in light of the mounting environmental crisis. This expansive digital realm offers tantalizing prospects, such as creating affordable spaces, fostering equity, and demolishing a myriad of barriers. Imagine a universe where, irrespective of your geographical, cultural, or economic background, you have the chance to experience, share, and grow. However, this horizon isn't without its shadows. As much as the metaverse promises an inclusive utopia, there's an underlying threat that we might inadvertently birth another form of digital inequality. In a world hurtling toward overwhelming environmental challenges, there may come a time when people are not just attracted but compelled to migrate into these digital expanses. This raises ethical quandaries about whether we're providing escape or merely transferring the locus of disparity. The Web 3.0 era we are navigating is a labyrinth of complexities. There's no singular metaverse but rather a dazzling array of digital universes, each with its own culture, rules, and possibilities. As we stand on the precipice of this vast unknown, it's crucial to recognize that its potential is as massive as its risks. With this understanding, we can hope to approach the metaverse with both wonder and wisdom, embracing the promise of the future while remaining vigilant about its inherent dangers.

Metaverse: A Digital Neverland?

In J. M. Barrie's (1904) timeless tale *Peter Pan*, we journey with children who navigate between the grounded realities of London and the fantastical realms of Neverland. It's a narrative that resonates with our innate human desire to straddle two worlds—the tangible and the imagined. In many ways, these adventures mirror our modern yearning for the metaverse. While we may inhabit a singular physical body, our spirits often long to exist simultaneously in two places, bridging the concrete and the virtual. The metaverse, much like Neverland, allows us to live between and beyond the real and the surreal, satisfying that profound wish to be a part of both realms. Just like the children in Neverland, we have this desire. It is now "realistically" fulfilled in the digital domain. Every Neverland, while mysterious, offers a balance of real and unreal. Our virtual and augmented realities, crafted for diverse dreams and needs, are simultaneously visible yet invisible to others, creating an ideal, individualized space that many are rushing into. However, while the whimsical charm of Neverland serves as a delightful analogy, the metaverse is not just a fairy-tale domain.

Predictions suggest that by 2026, a quarter of the global population will spend at least an hour daily in the metaverse, engaging in various activities. This expanding digital landscape integrates virtual reality experiences, spanning from gaming to business meetings, into a unified environment. A virtual economy is budding, fueled by digital currencies and non-fungible tokens (NFTs) that find their foundation on blockchains. Brands, ranging from athletic giants like Nike to food chains like Chipotle, are venturing into virtual stores and experiences, capitalizing on the vast potential of the metaverse (Murray, 2022). With investors spotlighting technological components and platforms pivotal to the metaverse experience, it's clear that many envision it as the next evolution of the internet. The corporate metaverse in particular is witnessing substantial momentum, catalyzed by the shift toward remote work and the promising prospects of virtual collaboration tools. While children in Neverland might engage in playful skirmishes with pirates or dance with fairies, the stakeholders in the metaverse navigate consequential actions that mold the future of industries, economies, and societies.

Metaverse: Future Communication Platform

The evolution of the metaverse is increasingly capturing the attention of technologists, scholars, and the general public alike. As it unfolds, the metaverse presents a complex landscape, encompassing a myriad of opportunities along with significant considerations that need to be thoughtfully explored. At its core, the metaverse offers a transformative mode through which to connect and engage with one another, especially with those that are far away. In virtual spaces, the traditional barriers of geography become almost meaningless. Relationships, both personal and professional, can be cultivated and maintained in ways that were previously unattainable, fostering a sense of closeness despite physical separation. The metaverse is more than just a technological marvel; it's a social revolution, opening up new avenues of communication and providing unprecedented freedom and choice for individuals. Through this expansive virtual landscape, people can transcend physical, social, and geographical constraints, presenting themselves and engaging with others in ways never before possible. With options ranging from avatars to virtual reality spaces, the metaverse enables richer personal expression, democratizes access, and fosters authentic connections. It represents a profound shift that acknowledges the complexity of human identity, allowing for nuanced and genuine interactions. The boundless possibilities of the metaverse invite us to redefine our understanding of communication and community in a world where more options genuinely mean more freedom. However, the potential benefits come with significant challenges to consider. Questions about accessibility, security, privacy, and ethical governance loom large. How will virtual spaces be regulated, and by whom? What safeguards must be in place to ensure that users' rights and dignities are respected? How will the metaverse include those who may lack the resources or capabilities to fully participate?

The trend toward embracing metaverse communication is accelerating, with broad consensus that these virtual interactions will continue to expand. It's not a fleeting phenomenon but a fundamental shift in how we understand and engage with technology and one another. As this landscape evolves, so too must our understanding and approach. The metaverse represents a remarkable opportunity for human interaction, full of potential yet demanding careful reflection and responsible stewardship. The balance between opportunity and consideration is delicate, and our journey into the metaverse must be undertaken with both optimism and caution. It's a new frontier that promises to reshape how we connect, but it's a path that must be navigated with wisdom and integrity.

Solving the Mobility Puzzle in the Metaverse

The current era of rapid technological advancement is marked by an intensified drive for global interconnectedness. This trend is bolstered by the appeal of increased mobility, as individuals and organizations alike seek enhanced contact and shared experiences on a global scale. However, this escalating mobility presents multifaceted challenges. On the environmental front, more frequent travel contributes to rising CO_2 emissions (Carrington, 2021), which is antithetical to global sustainability initiatives. From an economic perspective, the tangible costs linked to regular travel, in terms of both direct outlays and resource allocations, are significant. In this milieu, the metaverse, defined as a blended digital–physical reality, offers a promising solution. Within the metaverse, the constraints of geographical distance are largely mitigated, facilitating interactions that echo the richness of in-person exchanges without the accompanying environmental and economic repercussions. Essentially, the metaverse allows for the attenuation of spatial limitations, enabling intricate discussions or collaborative tasks without necessitating physical displacement. In the ongoing discourse surrounding mobility and its multifarious implications, the metaverse represents a pivotal paradigm. It suggests a future where the demands of global collaboration and interaction can be met without compromising environmental and economic sensibilities. As stakeholders across the world attempt to navigate the complexities of the "mobility puzzle," innovations like the metaverse could offer pathways toward sustainable interconnectedness.

The Space Revolution: From Tangible Rooms to Digital Realms

Amid our contemporary digital era, we're witnessing a profound shift in how we perceive and define space. Once confined strictly to tangible confines, our modern understanding of "space" comfortably embraces both the physical and virtual dimensions, inviting us to revisit and revise traditional notions. Consider the term "classroom." For generations, this term conjured up images of chalk-smeared blackboards, rows of wooden desks, and a palpable atmosphere of structured learning. Yet, in the present day, "classroom" has widened its scope of meaning. It now equally applies to virtual platforms where students from around the globe can gather, irrespective of time zones, for real-time or time-shifted education. This transformative trend is also mirrored in our evolving interpretation of "office." Tracing its etymological roots through the *OED*, we find that "office" first entered our language in the fourteenth century, primarily in ecclesiastical contexts. It was a descriptor for "A room, set of rooms, or building used as a place of business for non-manual work; a room or department for clerical or administrative tasks". As the centuries rolled on, the definition began to embrace a broader spectrum of professional environments, influenced by shifting work cultures and advancing technologies. Nowadays, "office" could refer to both a sleek downtown building and a digital workspace accessed from a personal computer. This evolution isn't merely semantic; it carries weighty societal implications.

The impact of physical lockdown and the shift to virtual interactions has evoked a diverse range of responses from people. While the longing for face-to-face interaction and the joy it brings remain palpable, this unique period has also presented us with an alternative model—one that holds promise for a more sustainable future. A report from 2019 highlights that American workers spend an average of almost 1 hour commuting to and from work each day, with even longer commutes prevalent in other regions (Burd et al., 2021). In India, the daily commute consumes approximately 2 hours, equivalent to 7 percent of an individual's day (ET Bureau, 2019). Amid the challenges of physical distancing, this new paradigm of virtual interactions has prompted individuals to reevaluate the necessity of extensive daily commuting. It has made us consider the potential for reducing the time spent on the road. While we all cherish the value of in-person connections, this experience has opened our eyes to the possibilities of adopting a more sustainable approach to daily interactions.

The evolving conceptions of workspaces have the potential to address urban population concentrations, a pressing concern for many global metropolises. For perspective, consider the dense population clusters in major cities like Tokyo, Delhi, Cairo, Mexico City, and Seoul. Tokyo alone, with its staggering population of over 37 million, represents approximately 29 percent of Japan's entire populace. The statistics for Delhi, Cairo, Mexico City, and Seoul tell similar stories, with population densities at approximately 2.3 percent, 9.8 percent, 17 percent, and 20 percent of their respective national totals (Galal, 2023; O'Neill, 2023; Yoon, 2023;). Yet, as our workspaces become increasingly decentralized, we're presented with an exciting possibility: the dispersion of workforces, reduced strain on urban infrastructure, and the alleviation of some of the perennial challenges associated with densely populated cities. Dense population concentrations bring about a slew of challenges. Notably, they exacerbate educational inequalities due to an uneven distribution of resources.

Nevertheless, as we stand on the cusp of an exciting digital frontier, there's a silver lining. The burgeoning metaverse, with its immersive virtual and augmented reality capabilities, heralds a new era of flexible working arrangements. People no longer need to be tethered to a physical location, reducing the stresses of commuting, and making it viable for individuals to reside in various regions while still being actively engaged in their professional pursuits. This decentralization has farther-reaching implications, too. With the metaverse's rise, we're looking at a future where work opportunities, educational resources, and cultural experiences are accessible irrespective of one's geographic location. This could lead to a more equitable distribution of resources, mitigating educational disparities, and significantly improving the quality of life for those outside urban epicenters. The changing tides are palpable even in our daily media consumption habits. Traditional newspapers, once the bedrock of daily information and journalism, face declining readerships. In the UK, expenditures on newspapers dipped from a robust 9.9 billion British pounds in 2005 to a mere 3 billion by 2021 (Watson, 2023b). By the first half of 2022, only two leading regional dailies in the UK reported circulations exceeding 25,000 (Watson, 2023b). The United States resonates with similar patterns, underscored by an increasing consumer pivot toward digital sources (Watson, 2023a). Regrettably, this has amplified the reach of less reliable news platforms. Around the world, a noticeable segment of readers has begun prioritizing influencers over seasoned journalists (Newman, 2023), a choice that sometimes compromises information accuracy.

Yet, it's not just news; our cultural and entertainment spaces are morphing as well. Conventional theaters and concert venues, while still vital cultural touchstones, find themselves sharing the stage with digital counterparts. Museums, too, are reaching beyond brick and mortar, enabling virtual tours and immersive online experiences that bring art and history to global audiences. In the commercial sector, the transformation is equally palpable. Once bustling shopping arcades and malls now share the consumer market with thriving e-commerce platforms. Virtual try-ons, augmented reality shopping guides, and direct digital channels have not just supplemented but often superseded traditional retail modalities, underscoring the deep digital integration reshaping our commercial behaviors. As our perceptions of traditional spaces continue to evolve, they herald innovative resolutions to longstanding societal challenges. This blending of the tangible with the digital captures the zeitgeist of our age, reflecting sweeping sociocultural and technological shifts while promising solutions to urban and infrastructural dilemmas that have been with us for ages.

VR Conferences: Think Earth and Environment

We should reconsider the notion that face-to-face meetings are the only or best option. As mentioned in earlier chapters, for some, the metaverse provides a safe and empowering space. Moreover, it can offer an economically efficient, inclusive, and environmentally friendly alternative. This paves the way for a more inclusive format. Recently, I had the opportunity to organize a VR-based international conference on "Language and AI in Asia: Possibilities and Challenges." It proved to be a resounding success, garnering positive feedback, particularly because of its highly interactive nature—quite distinct from most online conferences. In the instructional video available at https://tinyurl.com/2dky2zte, you can witness how we captured the essence of academic presentations while fostering socialization and networking. This conference approach was especially advantageous in encouraging participation from the global south, as the majority of our attendees hailed from these regions. Through this workshop, we managed to save a substantial £61,929.68 in costs and reduce CO_2 emissions by 44.68 tons, underscoring its significant economic and environmental impact. In terms of inclusivity, interactivity, and both economic and environmental considerations, VR conferencing represents the future. Of course, I am not suggesting that VR is the sole option, but rather a desirable one for international gatherings. We need to critically evaluate the merits of in-person and virtual meetings on a case-by-case basis, transcending the notion that face-to-face interactions are always the superior and optimal choice.

Environmental Concerns: The Metaverse as a Sustainable Solution

We thus find ourselves facing a dilemma as our need for connection and interaction in our multilingual, multicultural world continues to grow. In this era where constant interaction across regional boundaries has become the norm, we are confronted with a significant challenge: the carbon footprint. The greenhouse gas emissions stemming from commercial air travel are surpassing previous projections, intensifying the urgency for stronger actions by airline regulators as they prepare for an imminent summit. The United Nations aviation body predicts that carbon dioxide emissions from aeroplanes, a major contributor to greenhouse gases, will exceed 900 million metric tons in 2018 and potentially triple by 2050, if not sooner (International Civil Aviation Organization, 2019). This pressing issue is magnified as we witness the increasingly severe impacts of climate change on a yearly, and even daily, basis. Nevertheless, this does not imply that we should cease meeting altogether. Instead, we must reflect on how we can meet and interact in an environmentally friendly manner. This is where immersive communication emerges as an increasingly significant solution. However, before delving into the nature of immersive communication, it is vital to first compare online and offline meetings and find a delicate balance between the two. Evaluating this balance is no easy task, but it is a collective effort that humanity must undertake to ensure a sustainable future together.

The metaverse stands as not only a compelling substitute for physical spaces but is also intrinsically linked to the broader need for a sustainable, carbon-neutral environment, vital for the continued existence of humanity. In 2021, a comprehensive assessment conducted by the Joint Research Centre (JRC), the International Energy Agency (IEA), and the PBL Netherlands Environmental Assessment Agency revealed a disheartening surge of 5.3 percent in global fossil CO_2 emissions, drawing near to figures observed before the pandemic (Crippa, 2022). Although the European Union witnessed a 6.5 percent rise in emissions that year, it managed to sustain levels 5 percent below those of 2019, a testament to its consistent decline (Crippa, 2022). Remarkably, since 1990, the EU has reduced its fossil CO_2 emissions by 27.3 percent (Crippa, 2022). A wider observation indicates that many nations are significantly contributing to the worldwide CO_2 emissions. Every nation, irrespective of its emission figures, should be deeply cognizant of these ramifications and actively pursue remedial measures. The shared onus to confront and alleviate this environmental crisis has never been more crucial.

Employing data from the expansive Emission Database for Global Atmospheric Research (EDGAR), Crippa et al.'s analysis spans a variety of scenarios, ranging from wildfires to shifts in land use, encompassing insights from 208 countries. These revelations highlight the immediate demand for determined global measures. The unchecked ascent of CO_2 emissions threatens our planet's ecological equilibrium. This critical juncture accentuates the imperative to reevaluate our existing paradigms, particularly concerning international communication. As our planet contends with impending environmental predicaments, incorporating digital innovations such as metaverse conferencing platforms evolves from being a mark of advancement to an essential step toward a greener tomorrow. The evolving environmental context mandates nations, organizations, and individuals to reassess and adapt.

Impacts on Education: Accessibility and Affordability of the Metaverse

In an age where digital transformation is redefining boundaries, the metaverse emerges as a beacon of potential, especially in the realm of education. By transcending geographical limitations, the metaverse offers unparalleled opportunities to democratize access to knowledge. Whether it's global conferences, academic collaborations, or classroom interactions, this digital realm paves the way for more inclusive and cost-effective educational experiences. A collaborative report by several international organizations such as UNESCO, the World Bank, and UNICEF in 2022 announced that low- and middle-income countries are experiencing an educational crisis. Learning poverty has soared approximately 30 percent, meaning that seven out of ten 10-year-old children do not possess the literacy skills required to even comprehend basic written textsools like ChatGPT can aid these individuals, then we should embrace them. If ChatGPT-assisted interactive education can provide a high-quality, tailor-made education to underprivileged populations, we should welcome the opportunities. The central challenge is leveraging AI and the metaverse in ways that maximize benefits for all.

My late father's life story, a tale of unfulfilled academic aspirations due to financial constraints, serves as a powerful reminder of the value of education. This isn't just a tale from decades ago; it resonates with the present struggles of many around the world. While the affluent West is often the primary focus, we cannot overlook students in regions lacking resources and quality educators. It would be unjust to dismiss the potential of metaverse-based, AI-enhanced education without delving deep into its capabilities. Preliminary findings suggest that students, especially those grappling with motivation issues or anxiety, often thrive in such environments. This area warrants deeper exploration by educators and researchers. As we traverse the evolving landscape of the metaverse and AI in education, our mission is to harness their strengths, address challenges, and consistently prioritize the well-being and growth of students.

Legal Challenges in the Metaverse: Navigating Gray Areas

Navigating the complex labyrinth of the metaverse raises unprecedented legal challenges that our traditional, physical-world legal paradigms are ill-equipped to address. As an emerging frontier, the metaverse lacks set protocols and frameworks, leading to widespread confusion among users, developers, and even legal experts (Noordin & Nur, 2023). The very nature of this digital realm, absent of clear geographical boundaries, presents a foundational problem. When users from diverse countries, each backed by its unique legal system, interact in a shared digital environment, the question arises: Which jurisdiction's laws prevail during disputes? The intricate tapestry of the metaverse also weaves in the concept of digital assets, like virtual land, attire, or unique digital art. The recognition of these assets, their tangible value in the real world, and the rights associated with ownership, transfer, or inheritance are areas of law still in their infancy. And it isn't just about assets; the metaverse collects granular data about users' behaviors, preferences, and interactions (Fernandez & Hui, 2022). The protection, ownership, and potential monetization of this data throws up significant privacy issues.

Then there's the conundrum of crimes committed in the virtual world. Acts like cyberbullying, digital theft, or even instances where one's digital avatar is deleted or sabotaged present new challenges. These acts necessitate a reevaluation of our definitions of crime and justice. Traditional contractual obligations will also potentially undergo a transformation, with many metaverse interactions likely to be governed by digital or smart contracts. Their enforceability, breaches, and resolutions will need fresh legal perspectives. The multifaceted nature of individual representation in the virtual domain further complicates matters. As users may have multiple avatars or personas, pinning legal rights and responsibilities on these digital entities and linking them back to real-world identities is a legal maze.

With these challenges in mind, it's clear that a mere transplantation of our existing legal frameworks into the metaverse won't suffice. Instead, the laws governing this new realm must be fluid and adaptive, resonating with the rapid technological evolution we witness. Given the global nature of the metaverse, international collaboration in formulating legal frameworks seems not just beneficial but essential. Moreover, it's imperative that all stakeholders, from developers to users, play a role in shaping these laws, ensuring that the metaverse remains an equitable and inclusive space. As we venture further into this digital frontier, safeguarding digital rights becomes as crucial as protecting human rights in the tangible world. Crafting a robust legal framework for the metaverse isn't merely a requirement—it's a pressing obligation.

Metaverse as a Space of Equality and Potential

In today's society, where biases rooted in age, gender, ethnicity, and socioeconomic status frequently shape and limit interactions, the metaverse emerges as a transformative solution. This convergence of virtually enhanced physical reality, augmented reality, the internet, and persistent virtual spaces offers a compelling landscape to reshape our interactions. The metaverse acts as a powerful equalizer. In spaces where traditional barriers are especially pronounced, such as societies with entrenched social hierarchies like South Korea, the confines of the metaverse help dissolve these barriers. Within this expansive virtual realm, the norms of face-to-face interactions, which often restrict open expression based on an individual's social position, age, or gender, dissipate. Here, avatars, free from the bindings of real-world physicality, engage on an egalitarian plane. Attributes such as age, gender, or socioeconomic background become less influential in determining the course of an interaction. This egalitarian essence of the metaverse fosters a unique kind of communication where individuals are valued for their ideas, creativity, and virtual contributions rather than physical appearance or background. Imagine a virtual classroom or meeting where participants interact without the overt signals of real-world status or identity. Such settings can pave the way for more genuine, unbiased communication.

However, the metaverse isn't just about eliminating biases; it's also a bastion of opportunity. It represents a realm where innovation can flourish, unburdened by the typical constraints of the physical world. It's especially empowering for those who might feel marginalized or stifled by societal norms in traditional settings. Within the metaverse, individuals can delve deeper into aspects of their identity, passions, or talents they might hesitate to explore in the physical realm. However, as with all transformative technologies, the metaverse comes with its own set of challenges. While it holds the potential to revolutionize egalitarian interactions, there's a lurking danger that, without careful design and oversight, it could inadvertently mirror or even amplify the biases it aims to counteract. It's vital to ensure that these virtual spaces stay true to the promise of equality and inclusivity.

AI Ethics in the Metaverse

Developers working on AI and machine learning systems draw inspiration from various ethical frameworks and guidelines to shape their innovations. For instance, many reference the "Asilomar AI Principles." Crafted by a consortium of AI researchers and experts, these principles advocate for AI to be transparent, robust, and most importantly, in sync with human values (Gillis, 2023). Then there's the "AI Ethics Lab" framework, which delves deep into the ethical ramifications of AI, resting on four foundational pillars: fairness, reliability, safety, and explainability (Herranz, 2023). Not to be overlooked is the

"Partnership on AI," a nonprofit coalition dedicated to championing responsible AI practices (PAI Staff, 2023). Its membership boasts tech giants like Google, Microsoft, and IBM. However, as influential as these frameworks are, there's a pressing need for more cohesive transnational efforts. The reality is that when it comes to massive corporations, achieving true neutrality can be challenging. The immense influence wielded by these super-digital corporations can sometimes overshadow their ethical commitments. Hence, it's crucial to remain vigilant and be wary of the potential dangers tied to the unchecked power of these tech behemoths.

AI ethical codes are something that need to be considered on nation-wide and international scales urgently. In the UK, the GCHQ has unveiled a report titled "Pioneering a New National Security: The Ethics of Artificial Intelligence" (GCHQ, 2021). This significant document highlights the necessary steps that both the government and security agencies should embrace to ensure a safe transition into the AI-driven age. Within this report, GCHQ articulates the urgency of establishing a robust governance system for AI and data ethics. Such a system would incorporate best practices while actively seeking insights from a broad spectrum of external stakeholders. The goal is to set clear standards that developers can abide by, offering them both practical guidance and an educational program to aid in this process. Moreover, given the ever-evolving landscape of AI, the governance system itself must be adaptive. As AI technology advances and finds new applications, this system must be refined and expanded upon. To ensure this, GCHQ emphasizes a proactive approach, with continued engagement and a commitment to remain updated with the most recent developments. Above all, the report reinforces GCHQ's unwavering dedication to upholding privacy and human rights standards. The agency pledges to carry out thorough assessments, meticulously evaluating the necessity and balance of any potential infringements on privacy. This applies both when they harness operational data for the training and testing of AI tools and when these AI solutions are employed for analyzing operational datasets.

AI Regulations: "Better Late Than Never"

In the 2023 AI summit, the critical issue of implementing test-before-release procedures for AI models underwent extensive discussions. This significant initiative received support from ten countries, including the United Kingdom, the United States, Singapore, and Canada. Notably, China opted not to become a signatory. It's crucial to recognize that this agreement, while commendable, is just a part of the global picture. Like an atomic bomb, the potential risks of AI require a unified and worldwide effort to address effectively. Unless the entire world comes together to make a concerted move, the effectiveness of these regulations remains limited. In the digital world, where boundaries are less defined, enforcing such regulations poses significant challenges. While the move toward regulating AI is undoubtedly commendable, it's essential to recognize that it took a year to reach this milestone. ChatGPT, initially launched in November 2022, has made substantial progress during this period, amassing a user base of over 100 million active users, and attracting an astonishing 1.5 billion monthly visitors. As the old saying goes, "better late than never," but it also emphasizes the importance of continuous collaborative efforts and timely monitoring. Delays in addressing AI regulation can have far-reaching consequences, not only for individual nations but for all of humanity. Without doubt, AI technology will continue to advance at an incredible pace—though we must ask ourselves if we, the general public, are willing and able to keep up. The AI race has begun, and we are all participants.

Digital Capitalism: The Growing Influence of Mega-corporates

Digital capitalism is not a novel concept. As early as 2000, Daniel Schiller broached the subject in his book *Digital Capitalism: Networking the Global Market System*. However, what was once a burgeoning idea has now materialized into a palpable reality that casts its shadow across our digital interactions. In today's interconnected world, the hallmarks of digital capitalism are vividly apparent, with mega-sized corporations exerting substantial dominance over vast digital territories. Their influence permeates our interactions, habits, and even the dissemination of information. One of the primary tactics employed by these corporate behemoths is taking control of digital platforms. Initially, they captivate users with affordable or even complimentary subscriptions. Once embedded in the daily routines of users, and after fostering a sense of platform dependency, these corporations can pivot to more profit-centric models. While this elevates their revenue streams, it often burdens users with heightened costs or altered service structures. The cost, however, isn't just financial. In digital capitalism, data becomes a prime commodity. Mega-corporations have often been criticized for the potential mishandling or exploitation of user data, leading to concerns ranging from breaches of privacy to manipulative advertising tactics.

Moreover, these entities retain the power to alter operational mechanisms overnight. Such instances highlight the disproportionate power these corporations wield, often sidelining user preferences or concerns. The sheer dominance of these entities means they can instigate sweeping changes with little warning. A prime example is Twitter, a social media giant that boasted around 450 million monthly active users by 2022, marking an increase of over 40 percent since 2018 (Turner, 2023). However, by 2023, in a surprising turn of events, Elon Musk had transformed the platform, rebranding it from "Twitter" to "X Corp." The redirection of Twitter.com to X.com and the removal of Twitter's iconic bird logo from its San Francisco headquarters were emblematic of the abrupt changes these digital giants can impose. Decisions that can reshape online ecosystems are often made by a select few, leaving millions of users to adapt to these new realities.

This era's digital landscape, with its massive corporate authority, evokes Orwellian imagery of the "Big Brother." In George Orwell's *1984*, "Big Brother" stands as a foreboding symbol of absolute governmental control and surveillance, leaving citizens stripped of privacy (Orwell, 1949). Now, as we navigate the expansive digital realm of the metaverse, a new iteration of this omnipresent watcher emerges—not from a single, totalitarian state but from transnational corporations wielding enormous power in the virtual domain. These corporations, driven by profit and data, have the potential to assume a "Big Brother" role, continually monitoring users' interactions and leveraging this data for their benefit. With the metaverse offering unmatched convenience, personalization, and connectivity, users might find it tempting to embrace its offerings without reservations. However, this convenience may come at a hidden cost: potential control over our virtual lives and, by extension, our real-world perceptions and decisions. Just as Big Brother's omnipresence in *1984* symbolized an extreme invasion of individual rights and freedoms, the dominance of a few corporations in the metaverse could signify an analogous erosion of virtual autonomy. As we plunge deeper into this digital frontier, it becomes imperative to ensure that the metaverse remains a space of empowerment and creativity, rather than a playground for corporate surveillance and manipulation. What's more, as we tread further into this digital age, acknowledging the growing tentacles of digital capitalism becomes crucial. The need of the hour is to champion transparent, user-centric practices that counterbalance corporate profitability with individual rights and agency.

In today's swiftly evolving digital era, the principles of capitalism have seamlessly integrated with technology to create a new paradigm: digital capitalism. At its core, this system suggests that with more money, one can access better services, especially when various subscription levels come into play. However, this dynamic poses a significant risk. The inherent danger isn't a mere theoretical threat or a distant possibility. It's alarmingly imminent. As we increasingly rely on digital platforms and services, those with greater financial means can leverage premium offerings, while those with fewer resources find themselves at a notable disadvantage. This ever-widening gap between the "haves" and the "have-nots" is accentuated by the rapid growth of technology and digital services. If left unchecked, this growing chasm threatens to produce stark inequalities that could have lasting repercussions on society's fabric. It's essential to recognize and address this issue, ensuring that the digital future remains inclusive and doesn't inadvertently perpetuate existing socioeconomic disparities.

Danger of Techno-colonialism

Techno-colonialism is a multifaceted issue with implications that extend across the spheres of individual rights, governance, and social equality. The international scale of big tech companies complicates the process of regulation and accountability, making this a concern that transcends national borders. As society navigates the complexities of life in an increasingly digital world, the influence and control exerted by these tech giants will undoubtedly continue to be a subject of scrutiny and debate. In the digital era where technology advances at an exponential rate, the power dynamics between tech companies and users are increasingly becoming a topic of concern. The influence of big tech firms, often based in Western countries, on various aspects of life globally can be described as a form of "techno-colonialism." This narrative explores the potential implications of this phenomenon, which appears to be reshaping relationships between individuals, communities, and governing bodies.

The use of digital platforms for various aspects of daily life—from social interactions to healthcare—is more prevalent than ever. These platforms, provided largely by a handful of tech companies, collect vast amounts of data from their users. This data serves as a record of one's online activity and, by extension, is increasingly becoming a part of one's digital identity. The concentration of this form of personal information in the hands of a few corporations raises questions about who truly has control over these digital lives. The reach of these tech companies extends beyond the realm of data collection into the broader area of individual rights, including privacy and freedom of expression. Algorithms developed and employed by these platforms can determine what kind of news and information are most visible to users, thereby influencing public opinion and even electoral outcomes. Moreover, the international scale of these platforms often makes local governance a challenge, as local laws and customs can be easily bypassed or ignored. The relationship between big tech companies and their users is characterized by a significant imbalance in power. With extensive resources and sophisticated data analytics capabilities, these companies have the ability to influence both policy and public opinion. In contrast, the average user has limited means to contest these practices or to fully understand the extent of this influence. This dynamic can be seen as a digital version of traditional colonial relationships, where one party's significantly greater resources and capabilities are used to exert control over another.

The issue of techno-colonialism intersects with broader concerns about digital capitalism. Under this paradigm, control over digital resources—ranging from data to algorithms—translates into control over various social and economic dimensions. This convergence magnifies the risks associated with techno-colonialism, making it more than just an issue of data collection or privacy infringement; it becomes a vehicle for reinforcing existing social and economic inequalities.

AI Democracy in the Metaverse

In the midst of our rapidly digitizing world, it's essential to pause and reflect on a central question: Is our journey toward complete digital immersion in our best interest? Digital tools, including sophisticated chatbots like ChatGPT and predictive text software, have indeed reshaped the way we connect and communicate, offering unparalleled convenience and efficiency. Yet, the benefits might be counterbalanced by costs that aren't immediately apparent.

The post-Covid-19 era has further amplified this digital momentum. As cashless transactions grow in popularity, the humble banknote is almost rendered obsolete. But this digital push has its drawbacks. In many societies, the transition to digital methods, such as cashless payments, is no longer just an option but has become an imposed standard, leaving behind those who prefer or are more comfortable with traditional methods. Still, as the waves of digital transformation surge ahead, we must recognize and champion the rights of those who choose to anchor themselves in the nondigital domain. Whether driven by concerns of privacy, comfort with traditional methods, or a lack of digital accessibility and skills, their decision to remain nondigital is a right, not a defiance. The essence of the matter is not technological innovation in itself, but the sanctity of choice. Every individual's right to determine the extent and manner of their engagement with AI tools is paramount. This is the cornerstone of what we can term "AI democracy." Stephen Hawking's prescient observation that AI has the potential to be the end of mankind contrasts sharply with his reliance on a speech synthesizer (Cellan-Jones, 2014), underscoring the nuanced relationship we have with technology. As AI continues its march into our professional and personal spaces, it's essential to respect individual choices, even if they diverge from the mainstream. While there might be a transitional phase of adjustment, the hope remains that AI will usher in a brighter, more inclusive future for all. As we navigate these complexities, our language and lexicon are evolving too. Just as it feels inadequate to use the term "meet" for someone we've only engaged with online, so will the dawning age of AI and VR necessitate the birth of new vocabulary that encapsulates our shifting realities. As we stand at the crossroads of technological advancement and human values, it is our collective responsibility to ensure that the metaverse remains democratic, inclusive, and respectful of individual choices.

Right to Stay Nondigital

In today's increasingly digital world, there's a pertinent question we need to ask: Are all these strides toward digitalization unequivocally for our benefit? The rise of tools like chatbots, such as ChatGPT, and predictive text software, has revolutionized our interactions, promising a world of convenience at our fingertips. However, these conveniences might come at a price higher than we realize. Digital tools, by their very design, thrive on data collection. The constant feeding of personal data to these tools might create an illusion of control when, in reality, it could lead to manipulation. Our engagement with these tools could indirectly influence our behavior and possibly even endanger our diverse languages as major languages become more prevalent online. Take the case of text messaging, for instance. While it's undeniably convenient and efficient, it also opens a window for others to survey our online presence and activities. Features in apps like WhatsApp can reveal when a message has been read or when someone is typing, but does this level of transparency really serve our best interests? These considerations are far from straightforward. They call for active deliberation and thorough examination of the fine line between convenience and intrusion. In the wake of the Covid-19 pandemic, we've witnessed a significant uptick in digital adoption. Cashless transactions have become the new normal, sidelining cash to the brink of obsolescence in certain societies.

Amid this digital tide, we must consider those who may want to buck the trend. There are those among us who, for a variety of reasons, may choose to remain nondigital. One could be concerned about privacy, be more comfortable with traditional methods, or lack the resources or skills needed to transition to the digital world. The choice to stay nondigital is their right, and we must respect it. Such ethical considerations require serious, thoughtful conversations. As we sail through the digital era, it's critical that we champion the rights and preferences of every individual, ensuring that the balance between convenience and surveillance, and between inclusivity and coercion, isn't upset. The choice to stay nondigital should be a legitimate option rather than an act of defiance. As we steer the course of technological progression, we must ensure we're not leaving anyone unwillingly in its wake.

Metaverse Challenges for Human Communication in the Future

As we advance into an era where digital realms start to overshadow physical ones, pressing questions emerge about the future of our communication. Will the digital landscape fundamentally transform the ways we connect and converse? The traditional face-to-face interactions that have characterized human communication for millennia are now, in many ways, contending with digital modes of connection. People increasingly spend more time on social media than engaging in direct, in-person conversations. Communication can become more efficient in these spaces, often fostering solidarity and facilitating better expression. Could we envision a future where babies learn languages in the metaverse through AI-driven methods? Such possibilities might not be too distant.

Metaverse: Into the Unknown

As we conclude this chapter, it's crucial to take a moment and ruminate on the dynamic progression of the metaverse. At its core, the metaverse represents a swiftly expanding ecosystem, amalgamating digital dimensions that aim to mirror or even surpass our tangible realities. The speed of its metamorphosis is both astonishing and disconcerting. With every new dawn, unseen facets of this virtual cosmos manifest, each resonating with unparalleled potential. The promise is vast: from redefining how we interact socially to pioneering novel educational models, the metaverse is inexorably sculpting our shared destinies. Yet, venturing into uncharted territory always comes with risks. The metaverse, for all its allure, remains shrouded in mystery, a complex tapestry where digital and real-life threads are intricately woven together. This uncharted domain, where our existence oscillates between the palpable and the digital, poses a myriad of intricate ethical, philosophical, and technological conundrums. One of the primary challenges is the inherent decentralization of many metaverse platforms. This decentralization, while fostering innovation and reducing monopolistic control, also means that imposing any centralized ethical standards or regulatory frameworks becomes an uphill task. Censorship, in a conventional sense, becomes almost an unattainable goal. The lack of a centralized governing body can, therefore, lead to an environment where unethical behaviors can proliferate unchecked.

Finally, rapid technological leaps often eclipse our capacity as researchers to thoroughly analyze, comprehend, and relay timely interpretations. The lightning speed at which innovations manifest can leave the academic sphere gasping, endeavoring to contribute meaningfully to a domain that never stands still. This pace, however challenging, accentuates the gravity of our scholarly endeavors. It becomes imperative, now more than ever, to foster a healthy relationship with the metaverse, particularly for younger generations. We can't merely preach avoidance; we need to equip them with the tools to navigate this vast digital realm responsibly. Our continued exploration of the metaverse is not just a pursuit of knowledge but also a pressing obligation. Building a reliable, key platform that can be shared—noncommercial and beneficial—is of the essence. It's not about fostering fear or avoidance, but rather about constructing a dependable AI system for education, learning, and public services. That's the core objective. As academics, professionals, and denizens of the digital epoch, our task is to dig deeper, deciphering the multifarious implications of the metaverse in all its diverse contexts. It's a clarion call to ensure that as we journey deeper into this novel frontier, we remain grounded with foresight, prudence, and an unwavering pledge to the betterment of humanity.

References

Acharya, S., & Shukla, S. (2012). Mirror neurons: Enigma of the metaphysical modular brain. *Journal of Natural Science, Biology and Medicine, 3*(2), 118.

Ackerman, R., & Goldsmith, M. (2011). Metacognitive regulation of text learning: On screen versus on paper. *Journal of Experimental Psychology: Applied, 17*(1), 18–32. https://doi.org/10.1037/a0022086

ACODS UK. (2023, February 7). Web 3.0: The future of Internet. *Medium*. https://medium.com/@Acods/web-3-0-the-future-of-internet-75364ba1c8fa

Ahn, S. J. G., Le, A. M. T., & Bailenson, J. (2013). The effect of embodied experiences on self-other merging, attitude, and helping behavior. *Media Psychology, 16*(1), 7–38. https://doi.org/10.1080/15213269.2012.755877

Aiello, C., Bai, J., Schmidt, J., & Vilchynskyi, Y. (2022, June 13). Probing reality and myth in the metaverse. *McKinsey & Company*. https://www.mckinsey.com/industries/retail/our-insights/probing-reality-and-myth-in-the-metaverse

Akçayır, M., & Akçayır, G. (2017). Advantages and challenges associated with augmented reality for education: A systematic review of the literature. *Educational Research Review, 20*, 1–11. https://doi.org/10.1016/j.edurev.2016.11.002

Allely, S., Kesteven, S., & Leong, L. (2023, March 5). The 'phone lady' made a career out of teaching people how to speak on the phone. She's in high demand. *ABC News*. https://www.abc.net.au/news/2023-03-06/the-phone-lady-mary-jane-copps-phone-anxiety/102015450

Andembubtob, D. R., Keikhosrokiani, P., & Abdullah, N. L. (2023). A concise review on the concept of metaverse: Types, history, features, and technological perspectives. In P. Keikhosrokiani (Ed.), *Handbook of research on consumer behavioral analytics in metaverse and the adoption of a virtual world* (pp. 40–67). Hershey: IGI Global. https://doi.org/10.4018/978-1-6684-7029-9.ch003

Anderson, J., & Rainie, L. (2022, June 30). The metaverse in 2040. *Pew Research Center*. https://www.pewresearch.org/internet/2022/06/30/the-metaverse-in-2040/

Angelopoulos, S., & Merali, Y. (2015). Bridging the divide between virtual and embodied spaces: Exploring the effect of offline interactions on the sociability of participants of topic-specific online communities. In T. X. Bui & R. H. Sprague Jr (Eds.), *2015 48th Hawaii international conference on system sciences* (pp. 1994–2002). Piscataway: IEEE. https://doi.org/10.1109/HICSS.2015.239

Aoyagi, S., & Veliko, C. (2021, June 8). Preserve at-risk local languages. *UNESCO Multisectoral Regional Office in Bangkok*. https://bangkok.unesco.org/content/preserve-risk-local-languages

Apostolopoulos, J. G., Chou, P. A., Culbertson, B., Kalker, T., Trott, M. D., & Wee, S. (2012). The road to immersive communication. *Proceedings of the IEEE*, *100*(4), 974–990. https://doi.org/10.1109/JPROC.2011.2182069

Arguinbaev, M. (2017, October 30). Leeds University researchers worry about effect of VR headsets on young children. *The VR Soldier*. https://thevrsoldier.com/leeds-university-researchers-worry-about-effect-of-vr-headsets-on-young-children/

Armstrong, M. (2023, February 7). This chart shows how big the metaverse market could become. *World Economic Forum*. https://www.weforum.org/agenda/2023/02/chart-metaverse-market-growth-digital-economy/

Aubrey, J. S., Robb, M. B., Bailey, J., & Bailenson, J. (2018). *Virtual reality 101: What you need to know About kids and VR*. San Francisco: Common Sense. https://www.commonsensemedia.org/sites/default/files/research/report/csm_vr101_final_under5mb.pdf

Babel, M., & Russell, J. (2015). Expectations and speech intelligibility. *The Journal of the Acoustical Society of America*, *137*(5), 2823–2833. https://doi.org/10.1121/1.4919317

Bacca, J., Baldiris, S., Fabregat, R., Graf, S., & Kinshuk. (2014). Augmented reality trends in education: A systematic review of research and applications. *Journal of Educational Technology & Society*, *17*(4), 133–149. http://www.jstor.org/stable/jeductechsoci.17.4.133

Bailenson, J. (2018, January 17). Eight rules to help you stay safe in virtual reality. *Slate*. https://slate.com/technology/2018/01/eight-rules-to-help-you-stay-safe-in-virtual-reality.html

Baker, N. (2023, May 12). Online gaming statistics 2023. *Uswitch*. https://www.uswitch.com/broadband/studies/online-gaming-statistics/

Balis, J. (2022, July 27). How metaverse and Web 3 can create real value for your organization. *Ernst & Young*. https://www.ey.com/en_us/consulting/how-metaverse-and-web-3-can-create-real-value-for-your-organization

Banakou, D., Kishore, S., & Slater, M. (2018). Virtually being Einstein results in an improvement in cognitive task performance and a decrease in age bias. *Frontiers in Psychology*, *9*, 917. https://doi.org/10.3389/fpsyg.2018.00917

Barbot, B., & Kaufman, J. C. (2020). What makes immersive virtual reality the ultimate empathy machine? Discerning the underlying mechanisms of change. *Computers in Human Behavior*, *111*, 1–11. https://doi.org/10.1016/j.chb.2020.106431

Barnhart, B. (2023, April 28). Social media demographics to inform your brand's strategy in 2023. *Sprout Social*. https://sproutsocial.com/insights/new-social-media-demographics/

Barrie, J. M. (1904). *Peter Pan, or the boy who wouldn't grow up*. London: Hodder & Stoughton.

Bates, M. (2009, November 10). The mind is a mirror. *Scientific American*. https://www.scientificamerican.com/article/the-mind-is-a-mirror/

Bauerlein, M. (2009, September 4). Why Gen-Y Johnny can't read nonverbal cues. *Wall Street Journal*. https://www.wsj.com/articles/SB10001424052970203863204574348493483201758

Baumeister, R. F. (1998). The self. In D. T. Gilbert, S. T. Fiske, & G. Lindzey (Eds.), *The handbook of social psychology* (4th ed., pp. 680–740). Boston: McGraw-Hill.

Baumgartner, T., Speck, D., Wettstein, D., Masnari, O., Beeli, G., & Jäncke, L. (2008). Feeling present in arousing virtual reality worlds: Prefrontal brain regions differentially orchestrate presence experience in adults and children. *Frontiers in Human Neuroscience*, *2*, 1–12. https://doi.org/10.3389/neuro.09.008.2008

BBC. (2023, May 23). France bans short-haul flights to cut carbon emissions. *BBC*. https://www.bbc.com/news/world-europe-65687665

BBC News. (2023a, November 1). AI named word of the year by Collins dictionary. *BBC News*. https://www.bbc.co.uk/news/entertainment-arts-67271252

BBC News. (2023b, October 31). Most of our friends use AI in schoolwork. *BBC News*. https://www.bbc.co.uk/news/education-67236732

Berkman, M. I., & Akan, E. (2019). Presence and immersion in virtual reality. In N. Lee (Ed.), *Encyclopedia of computer graphics and games* (pp. 1–10). Cham: Springer. https://doi.org/10.1007/978-3-319-08234-9_162-1

Beveridge, C., & Lauron, S. (2023, January 26). 160+ social media statistics marketers need in 2023. *Hootsuite*. https://blog.hootsuite.com/social-media-statistics-for-social-media-managers/#Snapchat_statistics

Bianzino, N. M. (2022, September 4). How the metaverse could bring us closer to a sustainable reality. *VentureBeat*. https://venturebeat.com/virtual/how-the-metaverse-could-bring-us-closer-to-a-sustainable-reality/

Bickerton, J. (2023, June 30). Digital human created for Tour de France. *Sport Broadcast*. https://www.broadcastnow.co.uk/production/digital-human-created-for-tour-de-france/5183644.article

Blakemore, S.-J., & Mills, K. L. (2014). Is adolescence a sensitive period for sociocultural processing?. *Annual Review of Psychology*, *65*, 187–207. https://doi.org/10.1146/annurev-psych-010213-115202

Bloom, B. S. (1956). *Taxonomy of educational objectives: Cognitive and affective domains*. New York: David McKay.

Boeldt, D., McMahon, E., McFaul, M., & Greenleaf, W. (2019). Using virtual reality exposure therapy to enhance treatment of anxiety disorders: Identifying areas of clinical adoption and potential obstacles. *Frontiers in Psychiatry*, *10*, 1–6. https://doi.org/10.3389/fpsyt.2019.00773

Bonner, E., & Reinders, H. (2018). Augmented and virtual reality in the language classroom: Practical ideas. *Teaching English with Technology*, *18*(3), 33–53.

Born, M. P., & Taris, T. W. (2010). The impact of the wording of employment advertisements on students' inclination to apply for a job. *The Journal of Social Psychology*, *150*(5), 485–502. https://doi.org/10.1080/00224540903365422

Boss, S. (2008, June 12). Google lit trips: Bringing travel tales to life. *Edutopia*. https://www.edutopia.org/google-lit-trips-virtual-literature

Bower, M., Cram, A., & Groom, D. (2010). Blended reality: Issues and potentials in combining virtual worlds and face-to-face classes. In C. Steel (Ed.), *Proceedings of ASCILITE-- Australian Society for Computers in Learning in Tertiary Education Annual Conference 2010* (pp. 129–40). Brisbane: University of Queensland. https://www.ascilite.org/conferences/sydney10/procs/Bower-full.pdf

Bower, M., Howe, C., McCredie, N., Robinson, A., & Grover, D. (2014). Augmented reality in education – Cases, places and potentials. *Educational Media International*, *51*(1), 1–15. https://doi.org/10.1080/09523987.2014.889400

Boyles, B. (2017). *Virtual reality and augmented reality in education*. Center for Teaching Excellence, United States Military Academy. https://www.westpoint.edu/sites/default/files/inline-images/centers_research/center_for_teching_excellence/PDFs/mtp_project_papers/Boyles_17.pdf

Brandtzaeg, P. B., Skjuve, M. and Følstad, A. (2022). My AI friend: How users of a social chatbot understand their human–AI friendship. *Human Communication Research*, *48*(3), pp. 404–29. https://doi.org/10.1093/hcr/hqac008.

Brown Sr, M. A. (2020). Searching for answers to hybrid approaches in communication and learning environments. In M. Sarfraz (Ed.), *Innovative perspectives on interactive communication systems and technologies* (pp. 226–40). Hershey: IGI Global. https://doi.org/10.4018/978-1-7998-3355-0.ch011

Brownell, J. (2020). *The listening advantage: Outcomes and applications*. New York: Routledge. https://doi.org/10.4324/9781351118026

Buchholz, K. (2022, April 26). Where people spend the most & least time on social media. *Statista*. https://www.statista.com/chart/18983/time-spent-on-social-media/

Burd, C., Burrows, M., & McKenzie, B. (2021, March 18). *Travel time to work in the United States: 2019*. United States Census Bureau. https://www.census.gov/content/dam/Census/library/publications/2021/acs/acs-47.pdf

Cai, Y., Pan, Z., & Liu, M. (2022). Augmented reality technology in language learning: A meta-analysis. *Journal of Computer Assisted Learning*, *38*(4), 929–45. https://doi.org/10.1111/jcal.12661

Caldwell, V. (2021, November 18). 'I love her and see her as a real woman.' Meet a man who 'married' an artificial intelligence hologram. *CBC*. https://www.cbc.ca/documentaries/the-nature-of-things/i-love-her-and-see-her-as-a-real-woman-meet-a-man-who-married-an-artificial-intelligence-hologram-1.6253767

Carrillo, A. (2022, June 14). Nike-owned RTFKT Studios is bringing perfume to the metaverse. *Input*. https://www.inverse.com/input/style/rtfkt-studios-byredo-perfume-metaverse-nfts

Carrington, D. (2021, March 31). Elite minority of frequent flyers 'cause most of aviation's climate damage'. *The Guardian*. https://www.theguardian.com/world/2021/mar/31/elite-minority-frequent-flyers-aviation-climate-damage-flights

-environmental#:~:text=A%20global%20study%20reported%20by,fly%20at%20all%20that%20year

Case, W., & Epstein, A. (2021, November 8). The subscription economy has officially infiltrated gaming. *Morning Consult Pro.* https://pro.morningconsult.com/articles/the-subscription-economy-gaming

Cellan-Jones, R. (2014, December 2). Stephen Hawking warns artificial intelligence could end mankind. *BBC.* https://www.bbc.com/news/technology-30290540

Centre for Digital Business. (2020, February 5). Human conversations and digital humans ~ not just a pretty face. *Medium.* https://medium.com/@mariehjohnson/human-conversations-and-digital-humans-not-just-a-pretty-face-d5abfaaa2c5f

Challenger, Gray & Christmas. (2023, June 1). May 2023 layoffs jump on tech, retail, auto; YTD hiring lowest since 2016. *Challenger, Gray & Christmas, Inc.* https://www.challengergray.com/blog/may-2023-layoffs-jump-on-tech-retail-auto-ytd-hiring-lowest-since-2016/

Chegg. (2023, June). *What is the state of UK student mental health?* Center for Digital Learning. https://www.chegg.com/about/wp-content/uploads/2023/06/FINAL_Report-UK-Mental-Health_june.pdf

Chen, X., Siau, K., & Nah, F. F.-H. (2010). 3-D virtual world education: An empirical comparison with face-to-face classroom. In M. Lacity, S. March, & F. Niederman (Eds.), *Proceedings of the International Conference on Information Systems, ICIS 2010* (p. 260). St. Louis: Association for Information Systems. https://aisel.aisnet.org/icis2010_submissions/260/

Cho, J., Tom Dieck, M. C., & Jung, T. (2023). What is the metaverse? Challenges, opportunities, definition, and future research directions. In T. Jung, M. C. Tom Dieck, & S. M. C. Loureiro (Eds.), *Extended reality and metaverse: Immersive technology in times of crisis* (pp. 3–26). Cham: Springer. https://doi.org/10.1007/978-3-031-25390-4_1

Cho, Y.-J. (2023, February 8). Music video for MAVE:'s debut song surpasses 10 million views. *Korea JoongAng Daily.* https://koreajoongangdaily.joins.com/2023/02/08/entertainment/kpop/Korea-Kpop-Metaverse/20230208165411002.html

Chu, J., Qaisar, S., Shah, Z., & Jalil, A. (2021). Attention or distraction? The impact of mobile phone on users' psychological well-being. *Frontiers in Psychology, 12,* 1–12. https://doi.org/10.3389/fpsyg.2021.612127

Cline, E. (2011). *Ready player one.* New York: Broadway Books.

Clinton, V. (2019). Reading from paper compared to screens: A systematic review and meta-analysis. *Journal of Research in Reading, 42*(2), 288–325. https://doi.org/10.1111/1467-9817.12269

Coffee, P. (2023, May 12). Roblox criticized by children's advertising watchdog. *The Wall Street Journal.* https://www.wsj.com/articles/roblox-criticized-by-childrens-advertising-watchdog-8694a53b

Coulthard, M. (2004). Author identification, idiolect, and linguistic uniqueness. *Applied Linguistics, 25*(4), 431–47. https://doi.org/10.1093/applin/25.4.431

Cox, S. (2021, July 28). Why avatar style is the next fashion frontier for fashion brands. *We Are Social*. https://wearesocial.com/us/blog/2021/07/why-avatar-style-is-the-next-fashion-frontier-for-fashion-brands/

Crippa, M., Guizzardi, D., Banja, M., Solazzo, E., Muntean, M., Schaaf, E., Pagani, F., Monforti-Ferrario, F., Olivier, J. G. J., Quadrelli, R., Risquez Martin, A., Taghavi-Moharamli, P., Grassi, G., Rossi, S., Oom, D., Branco, A., San-Miguel, J., & Vignati, E. (2022). *CO2 emissions of all world countries: JRC/IEA/PBL 2022 report*. Luxembourg: Publications Office of the European Union. https://edgar.jrc.ec.europa.eu/booklet/CO2_emissions_of_all_world_countries_2022_report.pdf

Croes, E. A. J., & Antheunis, M. L. (2021). Perceived intimacy differences of daily online and offline interactions in people's social network. *Societies, 11*(1), 1–12. https://doi.org/10.3390/soc11010013

Cureton, D. (2023a, April 17). The role of the avatar in the metaverse. *XR Today*. https://www.xrtoday.com/virtual-reality/the-role-of-the-avatar-in-the-metaverse/

Cureton, D. (2023b, June 1). Virtual reality statistics to know in 2023. *XR Today*. https://www.xrtoday.com/virtual-reality/virtual-reality-statistics-to-know-in-2023/

Cuthbertson, A. (2022, May 23). 'The game is over': Google's DeepMind says it is on verge of achieving human-level AI. *The Independent*. https://www.independent.co.uk/tech/ai-deepmind-artificial-general-intelligence-b2080740.html

Dalim, C. S. C., Kolivand, H., Kadhim, H., Sunar, M. S., & Billinghurst, M. (2017). Factors influencing the acceptance of augmented reality in education: A review of the literature. *Journal of Computer Science, 13*(11), 581–9. https://doi.org/10.3844/jcssp.2017.581.589

Damio, S. M., & Ibrahim, Q. (2019). Virtual reality speaking application utilisation in combatting presentation apprehension. *Asian Journal of University Education, 15*(3), 235–44. http://doi.org/10.24191/ajue.v15i3.7802

Damio, S. M., & Ibrahim, Q. (2019). Virtual reality speaking application utilisation in combatting presentation apprehension. *Asian Journal of University Education, 15*(3), 235–44. http://doi.org/10.24191/ajue.v15i3.7802

data.ai Insights. (2022, January 12). The state of mobile in 2022: How to succeed in a mobile-first world as consumers spend 3.8 trillion hours on mobile devices. *data.ai*. https://www.data.ai/en/insights/market-data/state-of-mobile-2022/

Davis, A., Linvill, D. L., Hodges, L. F., Da Costa, A. F., & Lee, A. (2020). Virtual reality versus face-to-face practice: A study into situational apprehension and performance. *Communication Education, 69*(1), 70–84. https://doi.org/10.1080/03634523.2019.1684535

Dawkins, R. (1976). *The selfish gene*. New York: Oxford University Press.

de Looze, L. (2016). *The letter and the cosmos: How the alphabet has shaped the western view of the world*. Toronto: University of Toronto Press.

de Weck, O. L. (2022). *Technology roadmapping and development: A quantitative approach to the management of technology*. Cham: Springer. https://doi.org/10.1007/978-3-030-88346-1

Dean, B. (2023, March 27). Instagram demographic statistics: How many people use Instagram in 2023?. *Backlinko*. https://backlinko.com/instagram-users

Dede, C. J., Jacobson, J., & Richards, J. (2017). Introduction: Virtual, augmented, and mixed realities in education. In D. Liu, C. J. Dede, R. Huang, & J. Richards (Eds.), *Virtual, augmented, and mixed realities in education* (pp. 1–16). Singapore: Springer. https://doi.org/10.1007/978-981-10-5490-7_1

Delgado, P., & Salmerón, L. (2021). The inattentive on-screen reading: Reading medium affects attention and reading comprehension under time pressure. *Learning and Instruction, 71*, 1–13. https://doi.org/10.1016/j.learninstruc.2020.101396

Demirci, A., Karaburun, A., & Kılar, H. (2013). Using Google Earth as an educational tool in secondary school geography lessons. *International Research in Geographical and Environmental Education, 22*(4), 277–90. https://doi.org/10.1080/10382046.2013.846700

Diegmann, P., Schmidt-Kraepelin, M., Eynden, S., & Basten, D. (2015). Benefits of augmented reality in educational environments – A systematic literature review. In O. Thomas & F. Teuteberg (Eds.), *Wirtschaftsinformatik 2015 proceedings* (pp. 1542–56). Osnabrück: Osnabrück University. https://aisel.aisnet.org/cgi/viewcontent.cgi?article=1102&context=wi2015

Dixon, S. J. (2023, February 13). Number of social network users in selected countries in 2022 and 2027 (in millions). *Statista*. https://www.statista.com/statistics/278341/number-of-social-network-users-in-selected-countries/

Dixon, S. J. (2023a, February 24). Leading countries based on Facebook audience size as of January 2023 (in millions). *Statista*. https://www.statista.com/statistics/268136/top-15-countries-based-on-number-of-facebook-users/

Dixon, S. J. (2023b, May 3). Leading countries based on Snapchat audience size as of April 2023 (in millions). *Statista*. https://www.statista.com/statistics/315405/snapchat-user-region-distribution/

Dong, S.-H. (2021, August 12). Will AI-powered groups take over K-pop?. *The Korea Times*. https://www.koreatimes.co.kr/www/art/2021/06/398_310095.html

Dovchin, S. (2020). The psychological damages of linguistic racism and international students in Australia. *International Journal of Bilingual Education and Bilingualism, 23*(7), 804–18. https://doi.org/10.1080/13670050.2020.1759504

Dovchin, S. (2022, September 20). Linguistic racism can take a high toll on international students. *Times Higher Education*. https://www.timeshighereducation.com/campus/linguistic-racism-can-take-high-toll-international-students

Duarte, F. (2023, April 9). Average screen time for teens (2023). *Exploding Topics*. https://explodingtopics.com/blog/screen-time-for-teens

Eckert, P. (2012). Three waves of variation study: The emergence of meaning in the study of sociolinguistic variation. *Annual Review of Anthropology, 41*, 87–100. https://doi.org/10.1146/annurev-anthro-092611-145828

Eickelmann, B., & Vennemann, M. (2017). Teachers' attitudes and beliefs regarding ICT in teaching and learning in European countries. *European Educational Research Journal, 16*(6), 733–61. https://doi.org/10.1177/1474904117725899

El-Hadi, N. and Merino, D. (2023, March 30). Too many digital distractions are eroding our ability to read deeply, and here's how we can become aware of what's happening – Podcast. *The Conversation*. https://theconversation.com/too-many-digital-distractions-are-eroding-our-ability-to-read-deeply-and-heres-how-we-can-become-aware-of-whats-happening-podcast-202818

ET Bureau. (2019, September 3). Indians spend 7% of their day getting to their office. *The Economic Times*. https://economictimes.indiatimes.com/jobs/indians-spend-7-of-their-day-getting-to-their-office/articleshow/70954228.cms

Fan, M., Antle, A. N., & Warren, J. L. (2020). Augmented reality for early language learning: A systematic review of augmented reality application design, instructional strategies, and evaluation outcomes. *Journal of Educational Computing Research*, 58(6), 1059–1100. https://doi.org/10.1177/0735633120927489

Farber, D. (2007, January 9). Jobs: Today Apple is going to reinvent the phone. *ZDNet*. https://www.zdnet.com/article/jobs-today-apple-is-going-to-reinvent-the-phone/

Fernandez, C. B., & Hui, P. (2022, July). Life, the metaverse and everything: An overview of privacy, ethics, and governance in metaverse. In *2022 IEEE 42nd International Conference on Distributed Computing Systems Workshops (ICDCSW)* (pp. 272–7). Piscataway: IEEE. https://doi.org/10.1109/ICDCSW56584.2022.00058

Fiske, S. T., Cuddy, A. J. C., Glick, P., & Xu, J. (2002). A model of (often mixed) stereotype content: Competence and warmth respectively follow from perceived status and competition. *Journal of Personality and Social Psychology*, 82(6), 878–902. https://doi.org/10.1037/0022-3514.82.6.878

Fondo, M., & Jacobetty, P. (2020). Exploring affective barriers in virtual exchange: The telecollaborative foreign language anxiety scale. *Journal of Virtual Exchange*, 3(SI), 37–61. https://doi.org/10.21827/jve.3.36083

Galal, S. (2023, June 26). Total population of Egypt 2022, by governorate (in millions). *Statista*. https://www.statista.com/statistics/1229759/total-population-of-egypt-by-governorate/

Gao, R., Huang, S., Yao, Y., Liu, X., Zhou, Y., Zhang, S., Cai, S., Zuo, H., Zhan, Z., & Mo, L. (2022). Understanding *Zhongyong* using a *Zhongyong* approach: Re-examining the non-linear relationship between creativity and the Confucian doctrine of the mean. *Frontiers in Psychology*, 13, 1–15. https://doi.org/10.3389/fpsyg.2022.903411

Garcia, O., & Li, W. (2014). *Translanguaging: Language, bilingualism and education*. Basingstoke: Palgrave Macmillan. https://doi.org/10.1057/9781137385765

Garrido, L. E., Frías-Hiciano, M., Moreno-Jiménez, M., Cruz, G. N., García-Batista, Z. E., Guerra-Peña, K., & Medrano, L. A. (2022). Focusing on cybersickness: Pervasiveness, latent trajectories, susceptibility, and effects on the virtual reality experience. *Virtual Reality*, 26(4), 1347–71. https://doi.org/10.1007/s10055-022-00636-4

Garzón, J., Pavón, J., & Baldiris, S. (2019). Systematic review and meta-analysis of augmented reality in educational settings. *Virtual Reality*, 23(4), 447–59. https://doi.org/10.1007/s10055-019-00379-9

Gaucher, D., Friesen, J., & Kay, A. C. (2011). Evidence that gendered wording in job advertisements exists and sustains gender inequality. *Journal of Personality and Social Psychology, 101*(1), 109–28. https://doi.org/10.1037/a0022530

GCHQ. (2021, February 25). *Pioneering a new national security: The ethics of artificial intelligence.* https://www.gchq.gov.uk/files/GCHQAIPaper.pdf

Gendron, T. L., Welleford, E. A., Inker, J., & White, J. T. (2016). The language of ageism: Why we need to use words carefully. *The Gerontologist, 56*(6), 997–1006. https://doi.org/10.1093/geront/gnv066

Georgiev, D., & Ivanov, I. (2023, July 27). How much time do people spend on social media in 2023?. *Techjury.* https://techjury.net/blog/time-spent-on-social-media/

Gibbons, A. (2023, June 9). AI is moving too fast to regulate, security minister warns. *The Telegraph.* https://www.telegraph.co.uk/news/2023/06/09/security-minister-artificial-intelligence-regulation/

Gillath, O., McCall, C., Shaver, P. R., & Blascovich, J. (2008). What can virtual reality teach us about prosocial tendencies in real and virtual environments?. *Media Psychology, 11*(2), 259–82. https://doi.org/10.1080/15213260801906489

Gillis, A. S. (2023, March). *Asilomar AI principles.* TechTarget. https://www.techtarget.com/whatis/definition/Asilomar-AI-Principles

Giorgio, P., Jarvis, D., Auxier, B., Bobich, H., & Harwood, K. (2023, June 15). 2023 sports fan insights: The beginning of the immersive sports era. *Deloitte Insights.* https://www2.deloitte.com/uk/en/insights/industry/media-and-entertainment/immersive-sports-fandom.html

Godefridi, I., Suñer, F., Leblanc, C., & Meunier, F. (2021). Using virtual reality and peer feedback to reduce L2 speaking anxiety: An exploratory study. In N. Zoghlami, C. Brudermann, C. Sarré, M. Grosbois, L. Bradley, & S. Thouësny (Eds.), *CALL and professionalisation: Short papers from EUROCALL 2021* (pp. 100–5). Research-publishing.net. http://doi.org/10.14705/rpnet.2021.54.1316

Gupta, S. (2022, May 27). What is Web 3.0, and how does it impact digital marketers?. *Gartner.* https://www.gartner.com/en/digital-markets/insights/what-is-web-3-0

Gurman, M. (2023, June 6). Apple's $3,499 Vision Pro headset will test marketing might. *Bloomberg.* https://www.bloomberg.com/news/articles/2023-06-05/apple-debuts-vision-pro-headset-in-search-of-post-iphone-future#xj4y7vzkg

Hadid, A., Mannion, P., & Khoshnevisan, B. (2019). Augmented reality to the rescue of language learners. *Florida Journal of Educational Research, 57*(2), 81–9.

Hao, K. (2020, February 26). Robots that teach autistic kids social skills could help them develop. *MIT Technology Review.* https://www.technologyreview.com/2020/02/26/916719/ai-robots-teach-autistic-kids-social-skills-development/

Hein, R. M., Wienrich, C., & Latoschik, M. E. (2021). A systematic review of foreign language learning with immersive technologies (2001–2020). *AIMS Electronics and Electrical Engineering, 5*(2), 117–45. https://doi.org/10.3934/electreng.2021007

Herranz, M. (2023, July 14). The 4 pillars of Ethical AI – And why they're important to machine learning. *Pangeanic*. https://blog.pangeanic.com/the-4-pillars-of-ethical-ai-and-why-theyre-important-to-machine-learning

Hershenson, R. (1991, December 15). TV replacing teachers in classrooms. *The New York Times*. https://www.nytimes.com/1991/12/15/nyregion/tv-replacing-teachers-in-classrooms.html

Hill Holliday, Trilia, & Origin. (2019). *Meet Gen Z: The social generation. Part 2*. https://brand-news.it/wp-content/uploads/2019/10/2019-Gen-Z-Report.pdf

Hirose, A. (2022, April 5). 114 social media demographics that matter to marketers in 2023. *Hootsuite*. https://blog.hootsuite.com/social-media-demographics/#TikTok_demographics

Holmberg, K., & Huvila, I. (2008). Learning together apart: Distance education in a virtual world. *First Monday*, *13*(10). https://doi.org/10.5210/fm.v13i10.2178

Horvath, L. K., Merkel, E. F., Maass, A., & Sczesny, S. (2016). Does gender-fair language pay off? The social perception of professions from a cross-linguistic perspective. *Frontiers in Psychology*, *6*, 1–12. https://doi.org/10.3389/fpsyg.2015.02018

Howarth, J. (2023, January 13). Alarming average screen time statistics (2023). *Exploding Topics*. https://explodingtopics.com/blog/screen-time-stats

Howarth, J. (2023a, February 15). 7 key Gen Z trends for 2023. *Exploding Topics*. https://explodingtopics.com/blog/gen-z-trends

Howarth, J. (2023b, August 11). How many gamers are there? (New 2023 statistics). *Exploding Topics*. https://explodingtopics.com/blog/number-of-gamers

Horwitz, E. K., Horwitz, M. B., & Cope, J. (1986). Foreign language classroom anxiety. *The Modern Language Journal*, *70*(2), 125–32. https://doi.org/10.1111/j.1540-4781.1986.tb05256.x

Hua, Z., Lee, W., & Lyons, A. (2017). Polish shop(ping) as translanguaging space. *Social Semiotics*, *27*(4), 411–33. https://doi.org/10.1080/10350330.2017.1334390

Huang, X., Zou, D., Cheng, G., & Xie, H. (2021). A systematic review of AR and VR enhanced language learning. *Sustainability*, *13*(9), 1–28. https://doi.org/10.3390/su13094639

Huntington, C. (n.d.). Mirror neurons: Definition, function, & examples. *Berkeley Well-Being Institute*. http://berkeleywellbeing.com/mirror-neurons.html#:~:text=Mirror%20neurons%20may%20help%20explain,to%20do%20those%20things%20ourselves

International Civil Aviation Organization. (2019). *Item 15: Environmental protection – General provisions, aircraft noise and local air quality – Policy and standardization*. United Nations. https://www.icao.int/meetings/a40/documents/wp/wp_054_en.pdf

Iqbal, M. Z., Mangina, E., & Campbell, A. G. (2022). Current challenges and future research directions in augmented reality for education. *Multimodal Technologies and Interaction*, *6*(9), 1–29. https://doi.org/10.3390/mti6090075

Jabr, F. (2013, April 11). The reading brain in the digital age: The science of paper versus screens. *Scientific American*. https://www.scientificamerican.com/article/reading-paper-screens/

Jäckle, S. (2022). The carbon footprint of travelling to international academic conferences and options to minimise it. In K. Bjørkdahl & A. S. Franco Duharte (Eds.), *Academic flying and the means of communication* (pp. 19–52). Singapore: Palgrave Macmillan. https://doi.org/10.1007/978-981-16-4911-0_2

Jain, S., Thiagarajan, B., Shi, Z., Clabaugh, C., & Matarić, M. J. (2020). Modeling engagement in long-term, in-home socially assistive robot interventions for children with autism spectrum disorders. *Science Robotics, 5*(39), 1–9. https://doi.org/10.1126/scirobotics.aaz3791

James, W. (1890). *The principles of psychology*, Vol. 2. New York: Henry Holt and Company. https://doi.org/10.1037/11059-000

Jeffrie, N. (2022, December 8). The mobile gender gap in South Asia is now widening. *GSMA*. https://www.gsma.com/mobilefordevelopment/blog/the-mobile-gender-gap-in-south-asia-is-now-widening/

Jenkins, H. (2003, January 15). Transmedia storytelling. *MIT Technology Review*. https://www.technologyreview.com/2003/01/15/234540/transmedia-storytelling/

Jeong, H. (2023, January 24). [단독] "디지털 휴먼 故 박윤배 등장에 '회장님네 사람들' 제작진도 눈물" [Even the production team of 'Chairman's People' shed tears at the appearance of the late Park Yoon-bae as a digital human]. *Hankook Ilbo*. https://www.hankookilbo.com/News/Read/A2023011916560001682?did=tw

Jeong, L. (2020, October 23). BTS universe guide: K-drama youth tells the BTS members' story, plus Save Me webtoon, BT21 line friends characters and break the silence: The movie. *South China Morning Post*. https://www.scmp.com/magazines/style/celebrity/article/3106819/bts-universe-guide-k-drama-youth-tells-bts-members-story

Jiang, K. (2023, June 8). Council post: From science fiction to reality: How digital humans are forging new realities. *Forbes*. https://www.forbes.com/sites/forbestechcouncil/2023/06/08/from-science-fiction-to-reality-how-digital-humans-are-forging-new-realities/?sh=b6250c644fd4

Jie, Y.-E. (2022, August 24). [Feature] Is K-pop's 'universe' building still worth it? *The Korea Herald*. https://www.koreaherald.com/view.php?ud=20220824000684#:~:text=The%20universe%20may%20not%20be,%2Dgeneration%20K%2Dpop%20acts

JOLLY. (2022, September 20). *I secretly replaced my co-host with AI!* [Video]. Youtube. https://www.youtube.com/watch?v=EPaCcFqGLsg

Kahn, J. (2023, March 29). Elon Musk and Apple cofounder Steve Wozniak among over 1,100 who sign open letter calling for 6-month ban on creating powerful A.I. *Fortune*. https://fortune.com/2023/03/29/elon-musk-apple-steve-wozniak-over-1100-sign-open-letter-6-month-ban-creating-powerful-ai/

Kang, S. (2007). Disembodiment in online social interaction: Impact of online chat on social support and psychosocial well-being. *CyberPsychology & Behavior, 10*(3), 475–77. https://doi.org/10.1089/cpb.2006.9929

Kaplan-Rakowski, R., & Gruber, A. (2023). The impact of high-immersion virtual reality on foreign language anxiety when speaking in public. *SSRN*. https://doi.org/10.2139/ssrn.3882215

Karacan, C. G., & Akoğlu, K. (2021). Educational augmented reality technology for language learning and teaching: A comprehensive review. *Shanlax International Journal of Education, 9*(2), 68–79. https://doi.org/10.34293/education.v9i2.3715

Karsenti, P. T., Bugmann, J., & Gros, P.-P. (2017). *Transforming education with minecraft: Results of an exploratory study conducted with 118 elementary-school students*. Montréal: CRIFPE. https://education.minecraft.net/content/dam/education-edition/software-downloads/Minecraft_Research_Report_Karsenti-Bugmann_2017.pdf

Kemp, S. (2022, January 26). Digital 2022: Global overview report. *DataReportal*. https://datareportal.com/reports/digital-2022-global-overview-report

Kemp, S. (2022, January 26). Digital 2022: Time spent using connected tech continues to rise. *DataReportal*. https://datareportal.com/reports/digital-2022-time-spent-with-connected-tech#:~:text=Research%20from%20GWI%20reveals%20that,the%20internet%20across%20all%20devices

Kemp, S. (2023, July 20). Digital 2023 July global statshot report. *DataReportal*. https://datareportal.com/reports/digital-2023-july-global-statshot

Kent, C., Rechavi, A., & Rafaeli, S. (2019). The relationship between offline social capital and online learning interactions. *International Journal of Communication, 13*, 1186–1211.

Khoros Staff. (2023, May 3). Social media customer service: Importance and stats for 2023. *Khoros*. https://khoros.com/blog/social-media-customer-service-stats#most-users-communicate-with-brands-on-social-media

Kiaer, J. (2020). *Pragmatic particles: Findings from Asian languages*. London: Bloomsbury Academic.

Kiaer, J. (2023). *Emoji speak: Communication and behaviours on social media*. London: Bloomsbury Academic.

Kiaer, J. (2023a). *Emoji speak: Communication and behaviours on social media*. London: Bloomsbury Academic.

Kiaer, J. (2023b). *Multimodal communication in young multilingual children: Learning beyond words*. Bristol: Multilingual Matters. https://doi.org/10.21832/9781800413344

Kiaer, J. (2023c). *The future of syntax: Asian perspectives in an AI age*. London: Bloomsbury Academic.

Kiaer, J., Kim, L., Hua, Z., & Li, W. (2022). Tomorrow? Jayaji!(자야지): Translation as translanguaging in interviews with the director of Parasite. *Translation and Translanguaging in Multilingual Contexts, 8*(3), 260–84. https://doi.org/10.1075/ttmc.00094.kia

Kiaer, J., Morgan, J. M., & Choi, N. (2021). *Young children's foreign language anxiety: The case of South Korea*. Bristol: Multilingual Matters. https://doi.org/10.21832/9781800411616

Kilvington, D. (2021). The virtual stages of hate: Using Goffman's work to conceptualise the motivations for online hate. *Media, Culture & Society, 43*(2), 256–72. https://doi.org/10.1177/0163443720972318

Kim, N. J., & Kim, M. K. (2022). Teacher's perceptions of using an artificial intelligence-based educational tool for scientific writing. *Frontiers in Education, 7*, 1–13. https://doi.org/10.3389/feduc.2022.755914

Kim, R. (2022, December 30). Why so many of your favorite K-dramas are based on webtoons. *TIME*. https://time.com/6243447/rise-of-webtoons-k-dramas/

Kiran, H., & Tonogbanua, L. (2023, July 26). 9 shocking global gamer statistics in 2023. *Techjury*. https://techjury.net/blog/how-many-gamers-are-there/#:~:text=There%20are%20over%203.09%20billion,is%20valued%20at%20%24197.11%20billion

Kloss, K. (2022, September 8). Karlie Kloss: 'Fashion designers in the future won't just be sewing, they'll be coding'. *CNN*. https://edition.cnn.com/style/article/karlie-kloss-september-issues/index.html#:~:text=For%20example%2C%20tech%20is%20already,%2C%20they'll%20be%20coding

Kohli, D. (2023, March 3). Web 3.0 technologies: A game-changer for multiple industries. *The Times of India*. https://timesofindia.indiatimes.com/blogs/voices/web-3-0-technologies-a-game-changer-for-multiple-industries/

Koutromanos, G., Sofos, A., & Avraamidou, L. (2015). The use of augmented reality games in education: A review of the literature. *Educational Media International, 52*(4), 253–71. https://doi.org/10.1080/09523987.2015.1125988

LaFrance, A. (2016, February 5). A surprise twist in the mystery of the lost telegrams. *The Atlantic*. https://www.theatlantic.com/technology/archive/2016/02/telegrams-stop-found-stop-kinda/460161/

Laghi, F., Schneider, B. H., Vitoroulis, I., Coplan, R. J., Baiocco, R., Amichai-Hamburger, Y., Hudek, N., Koszycki, D., Miller, S., & Flament, M. (2013). Knowing when not to use the Internet: Shyness and adolescents' on-line and off-line interactions with friends. *Computers in Human Behavior, 29*(1), 51–7. https://doi.org/10.1016/j.chb.2012.07.015

Lear, C. A. (2020). The use of virtual reality to reduce L2 speaking anxiety. *Bulletin of Nagoya University of Foreign Studies (名古屋外国語大学論集), 6*, 147–69. https://doi.org/10.15073/00001410

Lee, J. Y., & Hemphill, A. (2022, December 12). K-pop: The rise of the virtual girl bands. *BBC*. https://www.bbc.com/news/world-asia-63827838

Lee, J.-L. (2023, January 2). Korean game industry revenue breaks 20 trillion won in 2021. *Korea JoongAng Daily*. https://koreajoongangdaily.joins.com/2023/01/02/business/industry/game-revenue/20230102151621398.html

Lee, K. (2012). Augmented reality in education and training. *TechTrends, 56*(2), 13–21. https://doi.org/10.1007/s11528-012-0559-3

Lee, P. S. N., Leung, L., Lo, V., Xiong, C., & Wu, T. (2011). Internet communication versus face-to-face interaction in quality of life. *Social Indicators Research, 100*(3), 375–89. https://doi.org/10.1007/s11205-010-9618-3

Lewandowski, M. (2010). Sociolects and registers–a contrastive analysis of two kinds of linguistic variation. *Investigationes Linguisticae, 20*, 60–79. https://doi.org/10.14746/il.2010.20.6

Lewis, G., Jones, B., & Baker, C. (2012). Translanguaging: Origins and development from school to street and beyond. *Educational Research and Evaluation: An International Journal on Theory and Practice, 18*(7), 641–54. https://doi.org/10.1080/13803611.2012.718488

Li, W. (2016). Multi-competence and the translanguaging instinct. In V. Cook & W. Li (Eds.), *The Cambridge handbook of linguistic multi-competence* (pp. 533–43). Cambridge: Cambridge University Press. https://doi.org/10.1017/CBO9781107425965.026

Lieberman, A., & Schroeder, J. (2020). Two social lives: How differences between online and offline interaction influence social outcomes. *Current Opinion in Psychology, 31*, 16–21. https://doi.org/10.1016/j.copsyc.2019.06.022

Lin, H., & Wang, H. (2014). Avatar creation in virtual worlds: Behaviors and motivations. *Computers in Human Behavior, 34*, 213–18. https://doi.org/10.1016/j.chb.2013.10.005

Ling, Y., Brinkman, W.-P., Nefs, H. T., Qu, C., & Heynderickx, I. (2012). Effects of stereoscopic viewing on presence, anxiety, and cybersickness in a virtual reality environment for public speaking. *Presence: Teleoperators and Virtual Environments, 21*(3), 254–67. https://doi.org/10.1162/PRES_a_00111

Linkedin. (n.d.). *How can parents and educators monitor and guide the use of VR headsets for kids?* https://www.linkedin.com/advice/1/how-can-parents-educators-monitor-guide-use-vr#:~:text=VR%20headsets%20are%20not%20suitable,for%20older%20teens%20or%20adults

Loewen, M. G. H., Burris, C. T., & Nacke, L. E. (2021). Me, myself, and not-I: Self-discrepancy type predicts avatar creation style. *Frontiers in Psychology, 11*, 1–7. https://doi.org/10.3389/fpsyg.2020.01902

Lyu, Y. (2019). *Using gamification and augmented reality to encourage Japanese second language students to speak English* [Master's thesis, KTH Royal Institute of Technology]. Digitala Vetenskapliga Arkivet. http://www.diva-portal.org/smash/record.jsf?pid=diva2%3A1416017&dswid=2925

Maas, M. J., & Hughes, J. M. (2020). Virtual, augmented and mixed reality in K-12 education: A review of the literature. *Technology, Pedagogy and Education, 29*(2), 231–49. https://doi.org/10.1080/1475939X.2020.1737210

Mahabarata, Y. (2021, October 29). Facebook changes name to Meta with vision of mastering metaverse: The universe of Ready Player One is realized. *VOI*. https://voi.id/en/bernas/98901

Majid, S. N. A., & Salam, A. R. (2021). A systematic review of augmented reality applications in language learning. *International Journal of Emerging Technologies in Learning, 16*(10), 18–34. https://doi.org/10.3991/ijet.v16i10.17273

Mangen, A., Walgermo, B. R., & Brønnick, K. (2013). Reading linear texts on paper versus computer screen: Effects on reading comprehension. *International Journal of Educational Research, 58*, 61–68. https://doi.org/10.1016/j.ijer.2012.12.002

MarketsandMarkets. (2020, August). Virtual reality market with COVID-19 impact analysis by offering (Hardware and software), technology, device type (Head-mounted display, gesture-tracking device), application (Consumer, commercial, enterprise, healthcare) and geography - Global forecast to 2025. https://www.marketsandmarkets.com/Market-Reports/reality-applications-market-458.html

Matthews, S. L., Uribe-Quevedo, A., & Theodorou, A. (2020, November). Rendering optimizations for virtual reality using eye-tracking. In *2020 22nd Symposium on Virtual and Augmented Reality (SVR)* (pp. 398–405). Piscataway: IEEE. https://doi.org/10.1109/SVR51698.2020.00066

Matviienko, A., Müller, F., Schmitz, M., Fendrich, M., & Mühlhäuser, M. (2022). Skyport: Investigating 3D teleportation methods in virtual environments. In S. Barbosa, C. Lampe, C. Appert, D. A. Shamma, S. Drucker, J. Williamson, & K. Yatani (Eds.), *CHI '22: Proceedings of the 2022 CHI Conference on human factors in computing systems* (pp. 1–11). New York: ACM. https://doi.org/10.1145/3491102.3501983

Mazzaferro, G. (2018). Translanguaging as everyday practice: An introduction. In G. Mazzaferro (Ed.), *Translanguaging as everyday practice* (pp. 1–12). Cham: Springer. https://doi.org/10.1007/978-3-319-94851-5_1

McClure, E. (2020). Escalating linguistic violence: From microaggressions to hate speech. In L. Freeman & J. W. Schroer (Eds.), *Microaggressions and philosophy* (pp. 121–45). https://doi.org/10.4324/9780429022470-6

McDowell, M. (2022, June 10). A perfume in the metaverse? Byredo and Rtfkt bet on visual "aura". *Vogue Business*. https://www.voguebusiness.com/technology/a-perfume-in-the-metaverse-byredo-and-rtfkt-bet-on-visual-aura

McNair, K. (2023, January 31). Studying abroad can cost over $16,000 per semester—Here's how students plan to pay for it. *CNBC*. https://www.cnbc.com/2023/01/31/how-college-students-afford-study-abroad.html

Mehrabian, A. (1972). *Nonverbal communication*. New York: Routledge. https://doi.org/10.4324/9781351308724

Melnick, K. (2022, March 8). New app brings real-time 3D holograms To iPhone. *VRScout*. https://vrscout.com/news/new-app-brings-real-time-3d-holograms-to-iphone/

Menegatti, M., & Rubini, M. (2017). Gender bias and sexism in language. In M. Powers (Ed.), *Oxford research encyclopedia of communication*. Oxford University Press. https://doi.org/10.1093/acrefore/9780190228613.013.470

Menegatti, M., Crocetti, E., & Rubini, M. (2017). Do gender and ethnicity make the difference? Linguistic evaluation bias in primary school. *Journal of Language and Social Psychology*, *36*(4), 415–37. https://doi.org/10.1177/0261927X17694980

Meseure, P., & Kheddar, A. (2011). Collision detection. In P. Fuchs, G. Moreau, & P. Guitton (Eds.), *Virtual reality: Concepts and technologies* (pp. 383–409). London: CRC Press. https://doi.org/10.1201/b11612-26

Miehlbradt, J., Cuturi, L. F., Zanchi, S., Gori, M., & Micera, S. (2021). Immersive virtual reality interferes with default head–trunk coordination strategies in young children. *Scientific Reports*, *11*(1), 1–13. https://doi.org/10.1038/s41598-021-96866-8

Miller, A. (2021, December 9). VR resolution, field of view and the science of the human eye. *AR Insider*. https://arinsider.co/2021/12/09/vr-resolution-field-of-view-and-the-science-of-the-human-eye/

Milmo, D. (2021, October 28). Enter the metaverse: The digital future Mark Zuckerberg is steering us toward. *The Guardian*. https://www.theguardian.com/technology/2021/oct/28/facebook-mark-zuckerberg-meta-metaverse

Mobiquity. (2020, November 16). *The rise in digital adoption among baby boomers*. https://f.hubspotusercontent10.net/hubfs/1868764/The%20Rise%20in%20Digital%20Adoption%20Among%20Baby%20Boomers.pdf

Molina, B. (2016, July 11). With 'Pokémon Go', augmented reality is having its moment. *USA Today*. https://www.usatoday.com/story/tech/gaming/2016/07/11/pokemon-go-augmented-reality/86950180/

Molla, R. (2019, October 29). Poor kids spend nearly 2 hours more on screens each day than rich kids. *Vox*. https://www.vox.com/recode/2019/10/29/20937870/kids-screentime-rich-poor-common-sense-media

Moody, R. (2023, March 15). Screen time statistics: Average screen time in US vs. the rest of the world. *Comparitech*. https://www.comparitech.com/tv-streaming/screen-time-statistics/

Moscatelli, S., Menegatti, M., Ellemers, N., Mariani, M. G., & Rubini, M. (2020). Men should be competent, women should have it all: Multiple criteria in the evaluation of female job candidates. *Sex Roles: A Journal of Research*, *83*(5–6), 269–88. https://doi.org/10.1007/s11199-019-01111-2

MTV News Staff. (2021, October 7). Aespa's 'savage' science fiction. *MTV*. https://www.mtv.com/news/xd7x51/aespa-ai-savage-interview

Murray, S. (2022, July 8). Big business dives into the metaverse. *UVA Today*. https://news.virginia.edu/content/big-business-dives-metaverse

Ndlovu, N. (2023, July 5). 120+ screen time statistics: Usage, effects, demographics and trends. *MarketSplash*. https://marketsplash.com/screentime-statistics/

Needleman, S. E. (2021, October 28). Facebook changes company name to Meta in focus on metaverse. *The Wall Street Journal*. https://www.wsj.com/articles/mark-zuckerberg-to-sketch-out-facebooks-metaverse-vision-11635413402

NEON. (2023, June 28). Warner Bros. Discovery Global Themed Entertainment and NEON partner to present Harry Potter: Visions of Magic – An all-new interactive art experience inspired by the Wizarding World. *Cityneon Holdings*. https://www.neonglobal.com/en/warner-bros-discovery-global-themed-entertainment-and-neon-partner-to-present-harry-potter-visions-of-magic-an-all-new-interactive-art-experience-inspired-by-the-wizarding-world/

Nesenbergs, K., Abolins, V., Ormanis, J., & Mednis, A. (2020). Use of augmented and virtual reality in remote higher education: A systematic umbrella review. *Education Sciences, 11*(1), 1–12. https://doi.org/10.3390/educsci11010008

Newman, E. J., & Schwarz, N. (2018). Good sound, good research: How audio quality influences perceptions of the research and researcher. *Science Communication, 40*(2), 246–57. https://doi.org/10.1177/1075547018759345

Newman, N. (2023, June 14). Overview and key findings of the 2023 digital news report. *Reuters Institute.* https://reutersinstitute.politics.ox.ac.uk/digital-news-report/2023/dnr-executive-summary

NGV Melbourne. (n.d.). *Shaun Gladwell.* https://www.ngv.vic.gov.au/melbourne-now/artists/shaun-gladwell/#:~:text=Gladwell%20simulates%20the%20experience%20of,cardiac%20arrest%20to%20brain%20death

Nichols, M. (2023, June 13). UN chief backs idea of global AI watchdog like nuclear agency. *Reuters.* https://www.reuters.com/technology/un-chief-backs-idea-global-ai-watchdog-like-nuclear-agency-2023-06-12/

Niiranen, J., Kiviruusu, O., Vornanen, R., Saarenpää-Heikkilä, O., & Paavonen, E. J. (2021). High-dose electronic media use in five-year-olds and its association with their psychosocial symptoms: A cohort study. *BMJ Open, 11*(3), 1–9. http://doi.org/10.1136/bmjopen-2020-040848

Nisiforou, E. A., Kosmas, P., & Vrasidas, C. (2021). Emergency remote teaching during COVID-19 pandemic: Lessons learned from Cyprus. *Educational Media International, 58*(2), 215–21. https://doi.org/10.1080/09523987.2021.1930484

Nobel Prize. (n.d.). *Dennis Gabor: Facts.* The Nobel Prize. Retrieved August 17, 2023, from https://www.nobelprize.org/prizes/physics/1971/gabor/facts/

Noordin, K. A., & Nur, N. M. (2023, April 4). A reality check on the metaverse. *The Edge Singapore.* https://www.theedgesingapore.com/digitaledge/focus/reality-check-metaverse

O'Neill, A. (2023, April 12). India: The ten largest cities in 2023 (in million inhabitants). *Statista.* https://www.statista.com/statistics/275378/largest-cities-in-india/

Ofcom. (2016, August 4). *Communications market report 2016.* https://www.ofcom.org.uk/__data/assets/pdf_file/0024/26826/cmr_uk_2016.pdf

Ofcom. (2022a, March 20). *Adults' media use and attitudes report 2022.* https://www.ofcom.org.uk/__data/assets/pdf_file/0020/234362/adults-media-use-and-attitudes-report-2022.pdf

Ofcom. (2022b, June 1). *Online nation 2022 report.* https://www.ofcom.org.uk/__data/assets/pdf_file/0023/238361/online-nation-2022-report.pdf

Ofcom. (2022c, August 17). *Media nations 2022: Interactive report.* https://www.ofcom.org.uk/research-and-data/tv-radio-and-on-demand/media-nations-reports/media-nations-2022/media-nations-2022-interactive-report

Ofcom. (2023, March 29). *Children and parents: Media use and attitudes.* https://www.ofcom.org.uk/__data/assets/pdf_file/0027/255852/childrens-media-use-and-attitudes-report-2023.pdf

Okdie, B. M., Guadagno, R. E., Bernieri, F. J., Geers, A. L., & Mclarney-Vesotski, A. R. (2011). Getting to know you: Face-to-face versus online interactions. *Computers in Human Behavior, 27*(1), 153–59. https://doi.org/10.1016/j.chb.2010.07.017

Oputu, E. (2020, September 16). The coronavirus pandemic has made communication more important than ever. *Temple Now*. https://news.temple.edu/news/2020-09-16/coronavirus-pandemic-has-made-communication-more-important-ever

Oranç, C., & Küntay, A. C. (2019). Learning from the real and the virtual worlds: Educational use of augmented reality in early childhood. *International Journal of Child-Computer Interaction, 21*, 104–111. https://doi.org/10.1016/j.ijcci.2019.06.002

Orwell, G. (1949). *1984*. London: Secker & Warburg.

Osler, L., & Zahavi, D. (2022). Sociality and embodiment: Online communication during and after Covid-19. *Foundations of Science*, 1–18. https://doi.org/10.1007/s10699-022-09861-1

PA Media. (2020, January 30). Most children own mobile phone by age of seven, study finds. *The Guardian*. https://www.theguardian.com/society/2020/jan/30/most-children-own-mobile-phone-by-age-of-seven-study-finds

Papanastasiou, G., Drigas, A., Skianis, C., Lytras, M., & Papanastasiou, E. (2019). Virtual and augmented reality effects on K-12, higher and tertiary education students' twenty-first century skills. *Virtual Reality, 23*(4), 425–36. https://doi.org/10.1007/s10055-018-0363-2

Park, M. (2020, February 14). South Korean mother given tearful VR reunion with deceased daughter. *Reuters*. https://www.reuters.com/article/us-southkorea-virtualreality-reunion-idUSKBN2081D6

Parmar, D., & Bickmore, T. (2020). Making it personal: Addressing individual audience members in oral presentations using augmented reality. *Proceedings of the ACM on Interactive, Mobile, Wearable and Ubiquitous Technologies, 4*(2), 1–22. https://doi.org/10.1145/3397336

Parmaxi, A. (2023). Virtual reality in language learning: A systematic review and implications for research and practice. *Interactive Learning Environments, 31*(1), 172–84. https://doi.org/10.1080/10494820.2020.1765392

Parmaxi, A., & Demetriou, A. A. (2020). Augmented reality in language learning: A state-of-the-art review of 2014–2019. *Journal of Computer Assisted Learning, 36*(6), 861–875.

Patrizio, A. (2023, July 28). Top social media statistics to check out in 2023. *TechTarget*. https://www.techtarget.com/whatis/feature/Top-social-media-statistics-to-check-out

PauseAI. (n.d.). PauseAI proposal. https://pauseai.info/proposal

Peck, T. C., Seinfeld, S., Aglioti, S. M., & Slater, M. (2013). Putting yourself in the skin of a black avatar reduces implicit racial bias. *Consciousness and Cognition, 22*(3), 779–87. https://doi.org/10.1016/j.concog.2013.04.016

Penn, P. (2022, June 10). Psychology explains why students aren't flocking back to campus. *Times Higher Education*. https://www.timeshighereducation.com/blog/psychology-explains-why-students-arent-flocking-back-campus

Perez, S. (2021, April 9). Consumers now average 4.2 hours per day in apps, up 30% from 2019. *TechCrunch*. https://techcrunch.com/2021/04/08/consumers-now-average-4-2-hours-per-day-in-apps-up-30-from-2019/

Petrosyan, A. (2023, February 24). Languages most frequently used for web content as of January 2023, by share of websites. *Statista*. https://www.statista.com/statistics/262946/most-common-languages-on-the-internet/

Pinker, S. (1994). *The language instinct: The new science of language and mind*. London: Allen Lane.

Poeschl, S. (2017). Virtual reality training for public speaking—A QUEST-VR framework validation. *Frontiers in ICT*, *4*, 1–13. https://doi.org/10.3389/fict.2017.00013

Portnoy, L. (2017, April 8). The neural technology of empathy and the virtual technology that will harness it. *Medium*. https://lportnoy.medium.com/the-neural-technology-of-empathy-and-the-virtual-technology-that-will-harness-it-2a5e89e11de3

Pratapa, H. (2021, January 10). Into the K-verse: Storytellers of K-pop. *offcultured*. https://offcultured.com/into-the-k-verse-storytellers-of-k-pop/

Pregowska, A., Masztalerz, K., Garlińska, M., & Osial, M. (2021). A worldwide journey through distance education – from the post office to virtual, augmented and mixed realities, and education during the COVID-19 pandemic. *Education Sciences*, *11*(3), 1–26. https://doi.org/10.3390/educsci11030118

Prensky, M. (2001). Digital natives, digital immigrants Part 1. *On the Horizon*, *9*(5), 1–6. https://doi.org/10.1108/10748120110424816

Pugliese, M., & Vesper, C. (2022). Digital joint action: Avatar-mediated social interaction in digital spaces. *Acta Psychologica*, *230*, 103758. https://doi.org/10.1016/j.actpsy.2022.103758

Punar Özçelik, N., Yangin Eksi, G., & Baturay, M. H. (2022). Augmented Reality (AR) in language learning: A principled review of 2017–2021. *Participatory Educational Research*, *9*(4), 131–152.

Punzo, G. (2022, October 21). Why Web 3.0 is relevant for content creators. *Forbes*. https://www.forbes.com/sites/forbestechcouncil/2022/10/21/why-web-30-is-relevant-for-content-creators/?sh=1c326c454e4d

PwC. (2019). Seeing is believing: How virtual reality and augmented reality are transforming business and the economy. https://www.pwccn.com/en/tmt/economic-impact-of-vr-ar.pdf

Qing, A., & Won, C. H. (2021, August 1). Social media use can trigger feelings of inferiority or inadequacy among young: Experts. *The Straits Times*. https://www.straitstimes.com/singapore/social-media-use-can-trigger-feelings-of-inferiority-or-inadequacy-among-young-experts

Qu, C., Ling, Y., Heynderickx, I., & Brinkman, W.-P. (2015). Virtual bystanders in a language lesson: Examining the effect of social evaluation, vicarious experience, cognitive consistency and praising on students' beliefs, self-efficacy and anxiety in a virtual reality environment. *PLoS One*, *10*(4), e0125279.

Racaniere, M. (2023, October 31). I wanted to become friends with an AI, here's what I learned. *Euronews*. https://www.euronews.com/next/2023/10/31/can-i-become-friends-with-an-ai-avatar-on-the-replika-app

Raj, T. I. (2021, July 1). Inside aespa's AI universe. *Nylon*. https://www.nylon.com/entertainment/aespa-ai-concept-explained-next-level-black-mamba

Ram, J. (2018, December 10). Second Life: Why does it matter in project management? *IPMA*. https://ipma.world/second-life-matter-project-management/

Roblox & The New School Parsons. (2022, October 31). *Metaverse fashion trends 2022*. https://blog.roblox.com/wp-content/uploads/2022/10/FINAL_2022-Metaverse-Fashion-Trends-report_Roblox-x-Parsons.pdf

Rosen, P. (2021, December 14). All museums will build a metaverse copy using NFTs, predicts contemporary art chief at world's largest museum. *Business Insider*. https://markets.businessinsider.com/news/currencies/museum-metaverse-twin-digital-art-nfts-state-hermitage-museum-russia-celestial-hermitage

Ruben, M. A., Stosic, M. D., Correale, J., & Blanch-Hartigan, D. (2021). Is technology enhancing or hindering interpersonal communication? a framework and preliminary results to examine the relationship between technology use and nonverbal decoding skill. *Frontiers in Psychology*, 11, 1–10. https://doi.org/10.3389/fpsyg.2020.611670

Ruby, D. (2023, April 1). 53+ gamers statistics for 2023 (Number of gamers & trends). *DemandSage*. https://www.demandsage.com/gamers-statistics/

Salumbides, A. (2022, October 12). Introducing Mesh avatars for Microsoft Teams in Private Preview. *Microsoft Tech Community*. https://techcommunity.microsoft.com/t5/microsoft-teams-blog/introducing-mesh-avatars-for-microsoft-teams-in-private-preview/ba-p/3646444

Satake, Y., Yamamoto, S., & Obari, H. (2021). Effects of virtual reality use on Japanese English learners' foreign language anxiety. In L. G. Chova, A. L. Martínez, & I. C. Torres (Eds.), *ICERI 2021 Conference Proceedings* (pp. 1234–40). Valencia: IATED Academy. https://doi.org/10.21125/iceri.2021.0358

Saunders, E. (2023, April 4). UK video games market value dipped by 5.6% in 2022. *BBC*. https://www.bbc.com/news/entertainment-arts-65175394

Scents of Wood. (2022, October). *Scents of Wood NFT subscription collection*. OpenSea. https://opensea.io/collection/scents-of-wood-nft-subscription-collection

Schiller, D. (2000). *Digital capitalism: Networking the global market system*. Cambridge: MIT Press.

Sczesny, S., Formanowicz, M., & Moser, F. (2016). Can gender-fair language reduce gender stereotyping and discrimination?. *Frontiers in Psychology*, 7, 1–11. https://doi.org/10.3389/fpsyg.2016.00025

Segovia, K. Y., & Bailenson, J. N. (2009). Virtually true: Children's acquisition of false memories in virtual reality. *Media Psychology*, 12(4), 371–93. https://doi.org/10.1080/15213260903287267

Seon, H.-G. (2023, January 3). S.Korea's gaming industry reaches $15.7 billion in scale. *The Korea Economic Daily Global Edition*. https://www.kedglobal.com/korean-games/newsView/ked202301030010

Shi, Z., Tang, T., & Yin, L. (2020). Construction of cognitive maps to improve reading performance by text signaling: Reading text on paper compared to on screen. *Frontiers in Psychology, 11*, 1–11. https://doi.org/10.3389/fpsyg.2020.571957

Shifman, L. (2013). *Memes in digital culture*. Cambridge: MIT Press. https://doi.org/10.7551/mitpress/9429.001.0001

Slater, M., Pertaub, D.-P., & Steed, A. (1999). Public speaking in virtual reality: Facing an audience of avatars. *IEEE Computer Graphics and Applications, 19*(2), 6–9. https://doi.org/10.1109/38.749116

Sprout Social. (2023, March 23). 50+ of the most important social media marketing statistics for 2023. *Sprout Social*. https://sproutsocial.com/insights/social-media-statistics/

Stahlberg, D., Braun, F., Irmen, L., & Sczesny, S. (2007). Representation of the sexes in language. In K. Fiedler (Ed.), *Social communication* (pp. 163–87). New York: Psychology Press.

Stamarski, C. S., & Son Hing, L. S. (2015). Gender inequalities in the workplace: The effects of organizational structures, processes, practices, and decision makers' sexism. *Frontiers in Psychology, 6*, 1400. https://doi.org/10.3389/fpsyg.2015.01400

Takac, M., Collett, J., Bloom, K. J., Conduit, R., Rehm, I., De Foe, A., & Cikajlo, I. (2019). Public speaking anxiety decreases within repeated virtual reality training sessions. *PLoS One, 14*(5), e0216288.

Tan, J. (2023, March 14). Online personal branding boosts employability for Gen Z. *HRM Asia*. https://hrmasia.com/gen-z-jobseekers-optimise-online-personal-brands-to-boost-employability/

Tarr, R. (2006). *Using "Google Earth" in the history classroom: History review (Bedford, England)*. London: History Today Ltd.

Thorpe, J. R. (2021, February 10). Why you shouldn't worry about Tiktok destroying your attention span. *Bustle*. https://www.bustle.com/wellness/tiktok-attention-span-brain-effects-experts

Thrasher, T. (2022). The impact of virtual reality on L2 French learners' language anxiety and oral comprehensibility: An exploratory study. *CALICO Journal, 39*(2), 219–38. https://doi.org/10.1558/cj.42198

Tibbetts, M., Epstein-Shuman, A., Leitao, M., & Kushlev, K. (2021). A week during COVID-19: Online social interactions are associated with greater connection and more stress. *Computers in Human Behavior Reports, 4*, 1–7. https://doi.org/10.1016/j.chbr.2021.100133

Tõke, T. (2019, August 26). Fortnite And Roblox are changing social media as we know it. *HackerNoon*. https://hackernoon.com/fortnite-and-roblox-are-changing-social-media-as-we-know-it-joc531pl

Trageser, C. (2022, February 23). What the metaverse might mean for kids. *National Geographic*. https://www.nationalgeographic.com/family/article/what-the-metaverse-might-mean-for-kids

Turner, A. (2023, August). How many users does Twitter have? *BankMyCell*. https://www.bankmycell.com/blog/how-many-users-does-twitter-have

University of Leeds. (2017, September 14). Is virtual reality bad for our health? The risks and opportunities of a technology revolution. *Medium*. https://medium.com/university-of-leeds/is-virtual-reality-bad-for-our-health-the-risks-and-opportunities-of-a-technology-revolution-31520e50820a

Veszelszki, Á. (2017). *Digilect: The impact of infocommunication technology on language*. Berlin: De Gruyter Saur.

Walsh, E. (2020, January 26). Dive, surf, or skim?. *Spark & Stitch Institute*. https://sparkandstitchinstitute.com/reading-comprehension-digital-age/#:~:text=We%20don't%20read%20the,less%20linear%2C%20more%20selective%20way

Wang, Z., Guo, Y., Wang, Y., Tu, Y.-F., & Liu, C. (2021). Technological solutions for sustainable development: Effects of a visual prompt scaffolding-based virtual reality approach on EFL learners' reading comprehension, learning attitude, motivation, and anxiety. *Sustainability*, *13*(24), 1–15. https://doi.org/10.3390/su132413977

Watson, A. (2023a, March 9). U.S. newspaper industry - Statistics & facts. *Statista*. https://www.statista.com/topics/994/newspapers/#topicOverview

Watson, A. (2023b, April 24). Newspaper industry in the United Kingdom - Statistics & facts. *Statista*. https://www.statista.com/topics/5932/newspaper-industry-uk/#topicOverview

Wei, L. (2011). Moment analysis and translanguaging space: Discursive construction of identities by multilingual Chinese youth in Britain. *Journal of Pragmatics*, *43*(5), 1222–35. https://doi.org/10.1016/j.pragma.2010.07.035

Weng, Q. (2012). *An introduction to contemporary remote sensing*. New York: McGraw-Hill.

Westcott, K., Arbanas, J., Arkenberg, C., Auxier, B., Loucks, J., & Downs, K. (2022, March 28). 2022 digital media trends, 16th edition: Toward the metaverse. *Deloitte Insights*. https://www2.deloitte.com/za/en/insights/industry/technology/digital-media-trends-consumption-habits-survey/summary.html

Westcott, K., Loucks, J., Ciampa, D., & Srivastava, S. (2019, June 10). Digital media trends survey. *Deloitte Insights*. https://www2.deloitte.com/xe/en/insights/industry/technology/digital-media-trends-consumption-habits-survey/trends-in-gaming-esports.html

Whitehead, M. (2010). *Language and literacy in the early years 0–7*. London: Sage.

Wiest, B. (2019, November 4). Millennials hate phone calls, and they have a point. *Forbes*. https://www.forbes.com/sites/briannawiest/2019/11/04/millennials-hate-phone-calls-they-have-a-point/

Wolf, M. (2008). *Proust and the squid: The story and Science of the reading brain*. Cambridge: Icon Books Ltd.

Wootton, C., & Bronstein, M. (2022, November 1). Insights from our '2022 Metaverse Fashion Trends' report. *Roblox Blog*. https://blog.roblox.com/2022/11/insights-from-our-2022-metaverse-fashion-trends-report/

World Health Organization. (2022, October 5). Physical activity. https://www.who.int/news-room/fact-sheets/detail/physical-activity#:~:text=For%201%20year%20olds%2C%20sedentary,1%20hour%3B%20less%20is%20better

Wuensch, K. L., Aziz, S., Ozan, E., Kishore, M., & Tabrizi, M. (2008). Pedagogical characteristics of online and face-to-face classes. *International Journal on E-Learning, 7*(3), 523–32.

Yadav, M., Sakib, M. N., Feng, K., Chaspari, T., & Behzadan, A. (2019). Virtual reality interfaces and population-specific models to mitigate public speaking anxiety. In *2019th International Conference on Affective Computing and Intelligent Interaction (ACII)* (pp. 269–75). Piscataway: IEEE. https://doi.org/10.1109/ACII.2019.8925509

Yim, H. (2023, March 14). Are they for real? South Korean girl band offers glimpse into metaverse. *Reuters*. https://www.reuters.com/lifestyle/are-they-real-south-korean-girl-band-offers-glimpse-into-metaverse-2023-03-14/

Yoon, S.-Y. (2019, March 28). K-pop groups create a whole new world: Extensive narratives allow artists to do more than just music. *Korea JoongAng Daily*. https://koreajoongangdaily.joins.com/2019/03/28/features/Kpop-groups-create-a-whole-new-world-Extensive-narratives-allow-artists-to-do-more-than-just-music/3061186.html

Yoon, L. (2023, August 8). Population density in South Korea in 2021, by city and province (in inhabitants per square kilometer). *Statista*. https://www.statista.com/statistics/1112322/south-korea-population-density-by-province/#:~:text=Population%20of%20Seoul,total%20population%20live%20in%20Seoul

York, J. (2022, November 8). How 'non-verbal communication' is going digital. *BBC*. https://www.bbc.com/worklife/article/20221104-how-non-verbal-communication-is-going-digital

York, J., Shibata, K., Tokutake, H., & Nakayama, H. (2021). Effect of SCMC on foreign language anxiety and learning experience: A comparison of voice, video, and VR-based oral interaction. *ReCALL, 33*(1), 49–70. https://doi.org/10.1017/S0958344020000154

Young, J. Y., & Stevens, M. (2023, January 29). Will the metaverse be entertaining? Ask South Korea. *The New York Times*. https://www.nytimes.com/2023/01/29/business/metaverse-k-pop-south-korea.html

Yuen, S. C.-Y., Yaoyuneyong, G., & Johnson, E. (2011). Augmented reality: An overview and five directions for AR in education. *Journal of Educational Technology Development and Exchange, 4*(1), 119–40. https://doi.org/10.18785/jetde.0401.10

Zarraonandia, T., Aedo, I., Díaz, P., & Montero, A. M. (2014). Augmented presentations: Supporting the communication in presentations by means of augmented reality. *International Journal of Human-Computer Interaction, 30*(10), 829–38. https://doi.org/10.1080/10447318.2014.927283

Zhang, X., & Venkatesh, V. (2013). Explaining employee job performance: The role of online and offline workplace communication networks. *MIS Quarterly*, *37*(3), 695–722. https://doi.org/10.25300/MISQ/2013/37.3.02

Zitron, E. (2023, May 8). RIP Metaverse. *Insider*. https://www.businessinsider.com/metaverse-dead-obituary-facebook-mark-zuckerberg-tech-fad-ai-chatgpt-2023-5

Index

2D metaverse
   advantages 159
   classroom 183–4
   communication, challenges 159–60
   in education 158–60
   to teach the Korean language 144
3D AI holographic characters by
     Gatebox 137
3D headsets 178
3D spatial overlay visualization 130, 175
3E model (efficiency, expressivity, and
     empathy) 99, 103

aerial photography 107
AESPA 54, 109
age bias 136
   in language use 31–2
ageism 29
AI, *see* artificial intelligence (AI)
AI-assisted learning 148, 193
AI-assisted self-study 148
AI capabilities 53–6, 84
   negative reactions toward digital
     humans 55
   threat of illegal activities 56
AI Ethics Lab framework 207
AI literacy
   and AI natives 63
   communication skills 62
   concepts 61
   cultivating critical thinking 61
   cybersecurity awareness 62
   digital navigation 61
   ethical considerations 61
   media literacy 62
   and screen literacy 60–2
   using AI tools 61
AI Safety Summit in 2023, UK's 82
AI tool (AISS) backed by GPT-2 149
AI voice modeling technology 53
Akçayýr, G. 162
Akçayýr, M. 162

Alexa 46, 51, 61, 95
AltspaceVR 36
Amazon Prime 116
Andembubtob, D. R. 9
Angelopoulos, S. 100
Antheunis, M. L. 100
Apple Vision Pro 1
AR, *see* augmented reality (AR)
AR in education
   advantages 162
   applications 163
   functions 162
   future 163–4
   gaming 163
   key factors 162–3
   objects modeling 163
   positive attitudes towards 164
   technological challenges 161
artificial general intelligence (AGI) 71–2
artificial intelligence (AI)
   covers 53–4
   digital companionship 52
   ethics 52
   friendship 53
   genie 73
   intervention 71–3
   models 82
   natives 62–3, 65
   normal 53
   as a partner 52
   symbol 73
   and VR, combination of 11
   VR and IP 98
Asilomar AI Principles 207
augmented reality (AR) 6–7, 10, 38, 40,
     43, 77–8, 109, 167, 201–2, 206
   blending real and virtual 7
   in language teaching 167
   lenient, with limited checks or
     censorship 38
   technologies 10
autism spectrum disorder (ASD) 156

Avatar (2009)   24
avatars
  animosity for diversity and empathy   36–7
  based interaction   159
  -based memes   54
  -based platforms   187
  characteristics   36
  choice   24–5
  communication   28–30
  customization   23–4, 28
  decorations   188
  evolution   36
  full-body avatars   36
  inclusive environment   30
  influence   136–7
  Mesh avatars   36
  open, equal exchange of ideas   29
  representation, cultural sensitivity in   27–8
  self-representation   25
  virtual identity   24–5
  VR with and without avatar   174

Baby Boomers   17, 39
Bacca, J.   162
Banakou, D.   136
A Bar at the Folies-Bergère (Édouard Manet's painting)   106
Barrie, J. M.   198
Bauerlein, M.   101
Baumeister, Roy   25
Bickmore, T.   130, 175
bi-literate reading   60
blockchain   12–13, 76, 124, 198
Bonner, E.   166–7
Born, M. P.   31, 62, 106
Bower, M.   161, 172
Boyles, B.   161
brands, in virtual stores   198
Brownell, J.   101
BTS   109, 189
Business Proposal (web novel)   115
bystander attitudes and beliefs   132

Cai, Y.   167
cashless transactions   212–13
censorship   38, 215
Center for Digital Learning, Chegg's   127

Challenger, Gray   72
ChatGPT   21, 23, 35, 40, 50, 54, 65, 69, 71–3, 129, 135, 144, 149, 205, 209, 212–13
Chen, G.   172
children
  AI into activities   73
  AR's impact on learning   165
  chat functionalities   56
  communication and gesture   69
  digital behavior   56–7
  digital exposure   164–5
  effective educational mobile applications   164–5
  high-dose electronic game use   164
  impact on children's brains   70
  language acquisition   69
  and language anxiety   128
  role of health professionals   164
  VR for   68–70
Childwise   64
Cho, J.   8
Chu, J.   160
classroom, metaverse   144
  AI-facilitated education   146
  human and AI educators   146
  quiz game   *145*
  whiteboard and multimedia resources   *145*
Cline, Ernest   10
cocreation   95
co-learning and metaverse   158
  age differences   158
  anonymity in online spaces   158
collective creativity   95
communication
  cues in   101
  digital   92
  emotional   100
  haptic   123–4
  in-person   101, 118
  intercultural   191–2
  interpersonal   52, 101
  multimodal   87
  nonverbal   80, 100–1, 174
  online   27, 98, 100–1, 118
communication apprehension (CA)   129
competitive games   57
conservation efforts   108

Console games 15
Constanza, Jean-Louis 113
Consumer Electronics Show (CES) 124
Copps, Mary Jane 14
Covid-19 2, 17, 35, 64, 95, 107, 146, 149–50, 169, 172–4, 179, 212, 214
creative storytelling 110
Crippa, M. 204
Croes, E. A. J. 100
crosstalk or background disturbances 102
cultural sensitivity in avatar representation 27–8
cute avatars 26–7
cuteness 26–7
cyberbullying 63, 206
cybersickness 18

Dalim, C. S. C. 162
Damio, S. M. 130, 174
Davis, A. 129, 133, 174
Dawkins, Richard 96
deep-reading 58
de Freitas, Dr. Nando 72
Demetriou, A. A. 171
Descript 5, 8, 19, 52, 54–5, 144, 200
Diegmann, P. 162
Digilect 80–3
digital assistants 46
digital avatars 9
digital capitalism
  and affordability 74–6
  monopolization of the metaverse 75
Digital Capitalism: Networking the Global Market System (Schiller) 209
digital communication 92, 102
digital companionship 52
digital currencies 198
digital dilemma 165
digital empathy 52
digital governance
  digital capitalism 75
  transnationalism 74
digital humans 23, 40, 46–51, 54–6, 72, 81, 135, 177
  AI and 3D animation technology 46
  applications and benefits 46–7
  avatars and emojis 50
  negative reactions toward 55
  talking in the metaverse 49–50
  and virtual humans 46–8
digital identities 9, 37–9
digital injustice 73–4
  affordability gap 74
  digital divides 73
  disparities in social media 73
Digital Media Trends survey (2022) 15
digital native 62, 101
Digital Natives, Digital Immigrants (Prensky) 62
digital-physical amalgamation 40
digital screen reading 57–60
digital space shift 77
  progression from large screens to mobile devices 77
  Web 3.0 (*see* Web 3.0)
digital theft 206
digital visualization 107
discovery-based learning 163, 166
Disney+ 116
distance education 172–4
distractions 56, 58, 63, 119, 155, 160
Dopamine 66

e-books 57, 113–14
Eickelmann, B. 149
Einstein virtual body 136–7
emergency response planning 108
Emission Database for Global Atmospheric Research (EDGAR) 204
emojis 21, 26, 36–8, 50, 54, 64, 70–1, 79, 84, 85, 87, 91, 99, 119
  birth of 96–7
  diversification and personalization 96
Emoji Speak 27, 38, 70, 86
emotional communication 100
emotions 9, 11, 14, 19, 27, 36, 39, 46, 48–9, 52, 55, 60, 66, 69, 84, 85, 87, 89, 93, 96, 100, 103, 105, 110, 121, 123, 137, 151, 153, 156, 166, 175, 184, 193
  avatars and emojis 119
  creativity and exploration 119
  differences between online and offline interaction 118
  emotional power of VR 137

interactions during social
  distancing 118
in metaverse 118–19
in online spaces 117–18
empathy 105–6
  and connectivity 11
  digital 52
  and sense of solidarity, Gen Z 120
English as an additional language
  (EAL) 134
English Medium Instruction (EMI) 92
entertainment
  digital 49
  evolution of 48
  gaming 41
  immersive 108–10
  Instagram 13, 18–19, 24, 41, 45, 63,
    66, 97, 140, 165
  transmedia storytelling 49, 109–10,
    115–16
  TV or video 4
  virtual 48
European Consortium for Political
  Research (ECPR) 2
EXO 110
expressivity 79, 99, 104, 118
extended reality (ER) 6–7, 85

Facebook 12–13, 41, 44, 63, 74, 97, 140,
  152, 165
face-to-face communication 2–3, 29–30,
  35–7, 52, 70, 85, 87, 93, 95, 99, 100,
  105, 118, 121, 139, 158–9, 172, 174,
  178, 180, 202–3, 207, 214
  blending of the virtual and physical
    worlds 173
  nonverbal cues 100
  Second Life's multimodal
    communication 172
  vs. virtual communication 172–3
  vs. VR with avatar *vs.* VR without
    avatar 174
Fan, M. 167, 170
fandom 48, 108–9
fear of being without phone 64
Field of view (FOV) 112
Fondo, M. 133
Foreign Language Classroom Anxiety
  Scales (FLCAS) 133

foreign language learning 161
  benefits 171
  with immersive technologies 170–2
  privacy concerns 170

Gabor, Dennis 122
gamers
  diverse demography 42
  gaming console 43
  PlayStation VR 43
  subscription-based gaming
    services 42
  "time limit feature" in games 139
  VR headsets 42
gamification 11, 71, 134, 150–1, 170, 188
  effect 150–1
  efficacy of the approach 151
  pathway for learning 150–1
gaming 14–16, 114–15
  AR gaming 163, 166
  avatar 23
  challenges 36
  demographic 42
  and education 65
  environments 7
  evolution of 114–15
  first mass-market application of the
    metaverse 43
  generational reach 14–16
  mobile games 15
  participatory model 114
  popularity 15
  rise of e-sports 115
  subscription-based gaming
    services 42
  360-degree views and immersive
    experiences 115
  Twitch 115
  virtual reality headsets 42–3
  VR gaming 138–9, 142
Gartner Inc. 46
Garzón, J. 161
Gather Town, case studies
  advantage in language
    education 189–90
  AI-assisted *vs.* traditional approaches,
    psychological stability 193
  boredom, non-English languages 194
  expressive 193

flexible resources   184–5
fun and reduced anxiety   192
gamified elements   188–9
hybrid methodology   177–8
inclusive classroom   187–8
integration of multimodal
    materials   184
intercultural communication   191–2
learners as teachers   190–1
metaverse classroom,
    advantages   179–80
metaverse in education   194–5
nonhierarchical environment   186–7
observations, dual learning
    environment   178–9
physical appearance concept   183
psychological comfort   180–1
real-time feedback   190
sense of community   183–4
socialization, opportunities for   189
teaching new scripts   189–90
technical glitches   194
transitioning to online
    modalities   183–4
uninterrupted collaboration   185–6
virtual and real-world instructional
    methods   178
visual engagement   182–3
Gen Alpha   21, 23, 64–5, 104
gender-based prejudices   28
gender bias   100
    in language use   30–1
gender-fair language   31
Gendron, T. L.   32
Gen MZ   3, 24
Gen X   15, 17, 39
Gen Z   4–5, 14–15, 17, 21, 23–4, 63–7,
    104, 146, 148
    and Alpha   65
    definition   64
    digital identities and personal
        convictions   38–9
    empathy and sense of solidarity   120
    tech-savvy   146
    time on social media   67
gesture   3, 51, 69, 87, 100
    gesture-tracking devices   107
    nonverbal   99, 101
    physical   183

Gladwell, Shaun   137
GOAT, AR application   134
Godefridi, I.   133
Google Earth
    AI-driven version of   108
    database of geospatial
        information   108
    digital visualization   107
Google Maps   20
Google Ngram Viewer   8
Google's Immersive View for
    Routes   20
GPS technology   20, 161
greenhouse gas emissions   203
Gruber, A.   133

Hadid, A.   167
haptic communication/technology
    123–4
Harry Potter: Visions of Magic   113
Hawking, Stephen   212, 213
Hein, R. M.   170
Hinglish in India   92
Holmberg, K.   172
holograms   122
holography   122
Horizon Worlds, Meta's   36
Hua, Z.   88
Huang, X.   171
Hughes, J. M.   163
Hugo, Victor   86
human-AI
    friendships   53
    interactions   51–2
    trust   52
humanness   47
humanoid robots   49, see also robots
Huvila, I.   172
Hybe Corporation   110
hybrid education
    limitations of metaverse   157–8
    virtual seminars   157

Ibrahim, Q.   130, 174
identity expression, fluidity in   39
idiolects   79–80, 90
    Third Wave of variation studies
        79–80
image words   91

immersive communication, *see also*
    artificial intelligence (AI)
  affordability   124–5
  challenges in VR spaces   98
  copyright   98
  definition   102
  description   102–3
  efficiency   103–4
  emojis   96–7
  empathy   105–6
  expressivity   104–5
  freedom to choose   103
  gesture   100
  impact on offline interactions   100–1
  intellectual property   98
  meme culture   97–8
  mute, power dynamics   102
  narratives   106
  online and offline   98–100
  technical reliability   124–5
immersive consumers   106–7
immersive experiences   7, 9, 16, 18, 46, 103, 107, 115, 119–21, 125, 128, 141
  fictional realms   113
  virtual realities (VR)   110–11, 119
immersive near-death experience   137–8
immersive virtual environments   6, 77
inclusive communication   93
in-person
  classroom setting   158
  communication   101, 118
  education   172
  nonverbal communicative cues   101
  and online classes, differences in pedagogical features   172
  social capital and virtual interactions, relationship   99
  vs. virtual   118, 160, 203
Instagram   13, 18–19, 24, 41, 45, 63, 66, 97, 140, 165
instant culture in digital life   18–19
International Atomic Energy Agency (IAEA)   82
International Energy Agency (IEA)   204
interpersonal communication   52, 101

Jacobetty, P.   133
James, William   25
Jones, Tricia   101

K-12   162–5
Kang, S.   101
Kaplan-Rakowski, R.   133
Karsenti, P. T.   151, 153–4
K/DA   54
Kent, C.   99
Kiaer, J.   62, 88
Kloss, Karlie   39
Koutromanos, G.   163
K-Pop
  fandom   48
  industry   109
  universe-building   109
Küntay, A. C.   164

Laghi, F.   118
Language Instinct (Pinker)   89
language teaching   178, 189, 194
  AR in   167–9
  context-driven and engaging experiences   169
  internet connectivity   168
  and VR   166–7
  "wow" or novelty effect   168
language use
  age bias in   31–2
  gender bias in   30–1
  racial bias in   33–4
Lear, C. A.   134
Lee, J. Y.   118, 163
letter-bound verbal communication   103
Li, W.   89
Lieberman, A.   118
lifelogging   14, 44–5, 49
Ling, Y.   175
linguistic(s)
  anarchy   91
  diversity   11, 34, 90
  inferiority complex   33
  racism   33–4
  signs of transformation   91
  in transition   51–2
LinkedIn   41, 69
live learning courses   16
Lorenz, Konrad   26
Lyu, Y.   134

Maas, M. J.   163
Majid, S. N. A.   171

masculine-associated words 30
MATSUKO, Slovakian software
    company 125
Mave (virtual K-pop group) 48
Mazzaferro, G. 88
McCartney, Paul 53
Mehrabian, Albert 100
memes 21, 38, 54, 70, 79, 84, 91, 96–8
    avatar-based memes 54
    challenges 97–8
    concept of 96
    creation of Memoji using
        smartphones 98
    form of AR 97
    multifaceted nature 97
Merali, Y. 100
Mesh avatars 36
metaverse
    activities 16
    advantage in language
        education 189–90
    AI democracy 212–13
    AI ethics 207–8
    for AI natives 63–5
    AI regulations 208–9
    anticipated and actual response
        times 19
    avatars (*see* avatars)
    blended digital-physical reality 200
    challenges for human
        communication 214
    characteristics of 9
    classroom 144–7
    and co-learning 158
    communication (*see* communication)
    *vs.*conventional platforms 154–6
    danger of techno-colonialism 211–12
    definition 6, 8–9
    digital avatars 9, 11
    digital capitalism 209–11
    diverse inhabitants of 40
    diversity 40–1, 84–5
    education 144–6, 204–5
    empathy and connectivity 11
    environmental concerns 203–4
    expansion of social media 13–14
    facial expressions and body
        language 99
    flexibility and hybridity 89–90
    gaming 11, 14–16 (*see also*
        gamification; gaming)
    grey areas 146–7
    hybrid model, fusion of online and
        offline modes 155
    identity expression 39
    instant culture 19
    issues of representation 11–12
    legal challenges 205–6
    mobility puzzle 199–200
    multisensory communication
        in 120–2
    OASIS 10
    power of speech 17
    psychology (*see* psychology of
        metaverse)
    relationships, personal and
        professional 198
    remote communication 84–5
    and robots 156–7
    screen time and socioeconomic
        factors 4
    skill set 89
    as a space of equality 206–7
    space revolution 200–2
    synergy, virtual and physical
        communication networks 99
    technical issues 18
    translanguaging competence 88–90
    virtual spaces 9, 20, 147–8
    VR conferences 202–3
    in Web 3.0 12–13
microaggressions 32–3
Microsoft Copilot 54
Microsoft Teams 2, 36, 102
Millennial cohort (Gen Y) 64
Millennials 15, 17, 21, 24, 38, 64, 66–7,
    146
Minecraft 43
    challenges 153–4
    in education 152–4
    in-game coding 151
    pedagogical use 153
    as transformative educational
        tool 151
mirror neurons 111
mixed reality (MR) 6–7, 146, 163–4, 173
mobility puzzle 199–200
moderation 35, 45, 68, 70, 84, 87–8

Mon-Williams, Mark 18, 68
Moscatelli, S. 30
multimodal(ity) 86–8, 114
   activities 191
   communication 87
   content 119
   dialogues 85
   information 104
   interactions 86, 113
   language usage 78
   media 104
   and multisensory forms 70
   Second Life's multimodal communication 172
   semiotic tools 89
   of translingual space 88
   value of 87
multisensory communication
   five primary senses 121
   visual cues 120
multiuser virtual environment (MUVE) 146
mute button, power dynamics of 102

Nesenbergs, K. 173
Netflix 42, 115–16
new tech anxiety 136
Niiranen, J. 164
nondigital
   activities 115
   choice to stay 213–14
non-fungible tokens (NFTs) 76, 198
nonhierarchical environment 186–7
nonverbal communication 80, 100–1, 174

OASIS (Ontologically Anthropocentric Sensory Immersive Simulation) 10
Okdie, B. M. 118
online
   communication 27, 98, 100–1, 118
   global communication during the pandemic 99
   meetings 2, 99
   and offline interactions 99
   tutorials or community forums 117
OpenAI 71
Oranç, C. 65, 164

Orwell, George 210
over-the-top (OTT) platforms 116
Özçelik, N., Punar 168–9

Papanastasiou, G. 165
paper books 57–8
   cognitive advantages 113
   deep, immersive reading 114
   specialized cognitive processes 113
Parmar, D. 130, 175
Parmaxi, A. 171
partnership on AI 207
PauseAI 81
PBL Netherlands Environmental Assessment Agency 204
personalized avatars 21
Personal Report of Public Speaking Anxiety, McCroskey's 133
phone phobia 14, 67
photos sharing with real-time filters 14
physical classroom setting 158, 172–3, 177–82, 184
physicality 60
Picturephone 102
Pinker, Steven 89
Pinterest 13, 41
Pioneering a New National Security: The Ethics of Artificial Intelligence (GCHQ) 208
Poeschl, S. 131, 175
Pokémon Go, immersive AR game 7, 20
post-Covid-19 era 212
PowerPoint presentations 155, 185
Pragmatic Particles (Kiaer) 99, 103
pre- ChatGPT era 73, 129
Pregowska, A. 173–4
pre-metaverse 16
psychology of metaverse
   anxiety, impact of VR 131–2, 134–6
   creative scholar's experience 141–2
   engagement and audience feedback 130
   ethical issues 137–8
   FLA, impact of VR 133–4
   generalization 129
   individual differences 131–2, 135–6
   motivation 136
   player's experience 138–9
   social anxiety 127–8

speed   128
student's experience   139–40
tool for creation   142
virtual bodies   136–7
VR as empowering space   128
VR in oral communication   129–30
public speaking   129, 131–4, 174–5
public speaking anxiety (PSA)   17, 127, 131–3, 174

Qu, C.   132
quick video snippets with virtual backgrounds   14

racial bias
   in language use   33–4
   role of avatars   36, 120
   VR help in reducing   67
reading   112–14
   "bi-literate" reading   60
   circuit   59
   deep-   58–9
   digital   57–9
   e-reading   58
   evolution of   112–14
   habits   59
   improvement with help of VPS-VR   136–7
   misreading   87
   online   58
   from paper   113–14
   for pleasure   57
   short-form   58
   skim-   57–8
   time-intensive deep reading processes   59–60
Reading Buddy (AR project)   167
Ready Player Me (Estonia's)   36
Ready Player One (Cline)   10
Reinders, H.   166–7
relationship-building   14, 99
remote communication   85, 155
Roblox   5, 17, 38–9, 43, 64, 150–2, 165
robots   49, 83–4, 156, 182
   autistic children, aid to   156
   in-home socially assistive robots   156
   and metaverse   156–7
RTFKT Clone X avatars   124
Ruben, M. A.   100

Salam, A. R.   171
Satake, Y.   134
satellite imagery   107
scent, recreation   124
Schiller, Daniel   209
Schroeder, J.   118
self, complexity of   25–6
Shifman, Limor   97
shopping   116–17
   immediacy of digital solutions   117
   product's online reputation   117
shorts   66, 104
   multitasking on social media   66
   super-short attention spans   66–7
Singlish in Singapore   92
singularity, concept of   71
Siri   46, 51, 61, 65, 95
skills training   129, 163, 166
skim-reading   57–8
Skype   101
Slater, M.   175
Snapchat   13, 41, 44, 66, 165
Snow Crash (Stephenson)   6
social anxiety
   kinds of anxiety   127
   physical symptoms   128
social chatbot Replika   53
social distancing   201–2
social media
   avatars on   24
   diversity   44
   educational tool   56
   expansion of   13–14
   feedback on   117
   Gen Z on   67
   language of   26–7
   and life logging   105
   multitasking on   66
   and role of dopamine   66
   sanctuary   35
   services   74
   spaces, expansion of   13–14
   users   41
   VR sanctuaries   35
social phobias   175
sound, proximity chat   123
South Korea's gaming industry   15
Spanglish in the United States   92
Spark & Stitch Institute   57

Spatial 36
Spielberg, S. 10
spontaneous interactions 157–8, 181, 183
Stephenson, Neal 6
Steve Jobs 1
storytelling 49, 109–10, 115–16
surveillance and security, issues of 35, 98, 210
synergy between AI and immersive technologies 108
Synthesia 54–5

Takac, M. 131, 175
Taris, T. W. 31
teachers/teaching
  to AI natives 65
  intervention 148, 157
  relationship with technology 148–50
  teacher-student dynamics 186–7
  teacher-student hierarchy 191
  teacher-student rapport 118, 155
techno-colonialism 211–12
Technology Acceptance Model 133
technophobia 149
Telecollaborative FLA Scale (T-FLAS) 133
telehealth appointments 16
text-to-voice chat 123
Thrasher, T. 135
Tibbetts, M. 118
TikTok 12, 24, 41, 63, 66, 77, 110, 116, 165
Tour de France 40
Tour de France Femmes 40
translanguaging 21
  act of 90
  competence 88–90
  instinct 89
  metaverse as 88
transverse, act of 19–20
Trump, Donald 110
Tugendhat, Tom 81
Tulsa rally 110
twenty-first-century skill 165–6
  connecting physical body with virtual 166–7
  effects and limitations of VR/AR 166
  tools for creating AR content 166
Twitter 12, 32, 66, 97, 165, 210

unity 25, 74, 159
urban planning 108
"use it or lose it" principle 60

Venkatesh, V. 99
Vennemann, M. 149
videoconferencing 101, 134
virtual humans 21, 46–7, 50–1, 54, 68, 137
virtual idols 48, 54
virtual immersion 86, 169
virtual meetings 3, 147, 203
virtual migration 86, 95
virtual reality (VR) 6–7, 98
  advantages 67
  and AI, combination of 11
  anxiety, reducing 67
  and AR tools 173–4
  collaborative activities 34
  definition 7
  exposure therapy 131
  eye-tracking 112
  Field of view (FOV) 112
  foreign language learning 161
  haptics 112
  instant messaging 67
  and IP 98
  locomotion in VR 112
  metaverse-like environment 34–5
  sanctuary 34–5
  time limits 68
  virtual worlds 35
  vocabulary 111–12
  for young children (*see* children)
virtual reality exposure therapy (VRET) 131
virtual shopping assistants 50
virtual therapy
  3D spatial overlay visualization 175
  VR and AR in public speaking 174–5
virtual *vs.* in-person communication 118
visual senses 122
von Ebers 132, 175
VR with visual prompt scaffolding (VPS-VR) 136
VTubers 49

Wang, Z. 136
Web 3.0 95, 116, 134, 197
  dialogue and smarter searches 13
  and diversity 77–9

era 197
fusion of diverse communication methods 79
images in Google search bar 78
metaverse in 12–13
user-centric paradigm 12
Wei, L. 88
Wiest, Brianna 67
workspaces, conceptions of 201
world-building 109
World Health Organization (WHO) 2, 68
World Wide Developer Conference (WWDC) 1

X.com 210
X Corp 210

Yadav, M. 131, 175
"year-abroad" language studies, concept of 169–70
York, J. 101, 135
Young, Jordan 53
Yuen, S. C.-Y. 166

Zae-In 51
Zarraonandia, T. 175
Zhang, X. 99
zhongyong (Doctrine of the Mean) 87–8
Zoom 2, 93, 95, 101–2, 134, 154–6, 159, 165, 182–5
    fatigue 156
Zuckerberg, M. 8

www.ingramcontent.com/pod-product-compliance
Lightning Source LLC
Chambersburg PA
CBHW071821300426
44116CB00009B/1394